Breakthroughs Mathematics and Problem-Solving Skills

BOOK 1

ROBERT MITCHELL

Project Editor
Ellen Carley Frechette

CB
CONTEMPORARY BOOKS

CHICAGO · NEW YORK

Material in this book also appears in Contemporary's
Pre-GED Mathematics and Problem-Solving Skills, Book 1.
Copyright © 1987 by Robert P. Mitchell.

Published by Contemporary Books, Inc.
180 North Michigan Avenue, Chicago, Illinois 60601
Manufactured in the United States of America
International Standard Book Number: 0-8092-4236-2

Published simultaneously in Canada by
Beaverbooks, Ltd.
195 Allstate Parkway
Valleywood Business Park
Markham, Ontario L3R 4T8
Canada

Consultant/Field Tester
Vivian Osth

Editorial Director
Caren Van Slyke

Editorial
Ann Upperco
Sarah Schmidt
Christine M. Benton

Production Editor
J. D. Fairbanks

Illustrator
Rosemary Morrissey-Herzberg

Art & Production
Princess Louise El
Lois Koehler
Marilyn Vevang

Cover Design
Lois Koehler

Typography
J•B Typesetting

Cover illustration by David Lee Csicsko
Photographed by Scott Simms

Breakthroughs in Mathematics and Problem-Solving Skills

BOOK 1

Contents

To The Instructor

Contemporary's new *Breakthroughs in Mathematics and Problem-Solving Skills* is a two-book series that's specially designed to prepare students in essential arithmetic and problem-solving skills.

Breakthroughs in Math, Book 1 instructs students in the skills of whole number arithmetic. Students learn to add, subtract, multiply, and divide, and to apply these arithmetic skills in a variety of word problems. Because of the growing importance of word problems, Book 1 carefully develops word-problem skills to complement a student's progress in computational skills.

The companion book, *Breakthroughs in Mathematics and Problem-Solving Skills, Book 2*, instructs students in the use of decimals, fractions, and percents. Book 2 begins with a general review of problem-solving skills and then further develops a student's competence in these skills while introducing the new computational topics.

Both Book 1 and Book 2 have several special features:

- "Skill Builders" are used throughout to help students make the transition from explanations to actual problem solving. "Skill Builders" are rows of specially-designed, partially-worked problems that the student completes before beginning practice exercises.

- Instruction in problem-solving skills is integrated throughout the text. Lessons in which these skills are emphasized are highlighted. Even when students can master computational skills, they often have difficulty in solving word problems. Carefully sequenced reasoning activities help students to build their problem-solving abilities.

- A special chapter, called "Special Topics In Math," is the final chapter in each book. In this chapter, students are introduced to special applications of computation and problem-solving skills. An example from Book 1 is Measurement; an example from Book 2 is Data Analysis.

- Practice tests are used throughout. An "Overview of Skills" begins each book, and a post-test concludes the student's study. Computation reviews are placed at the end of each chapter to enable students to reinforce computation skills before moving on to each new topic.

- A complete answer key is placed at the end of each book. In addition, solutions are worked out for all word problems so that students can see a step-by-step process to solve the problem correctly. This insures maximum teaching flexibility for the instructor and the possibility of immediate feedback for the students.

Equally important as quality learning materials is the attitude that students bring to the study of mathematics. Though good books help, it is the instructor who can do the most to help students learn the study habits that can make math an enjoyable and successful experience. One technique that is particularly helpful is to teach students to leave each day's study of math with a positive feeling. A good way for students to do this is to develop the habit of reviewing mastered problems as the last step before putting a math book away. By reworking a few known problems successfully, a student both reinforces concepts learned that day, and reminds himself or herself of the progress being made.

Leaving math each day on such a positive note builds positive memories about the study of math, and that may well be the most important factor that determines student success.

To The Student

Welcome to *Breakthroughs in Mathematics, Book 1.*

This book is designed to provide you with the skills of whole number arithmetic. You will learn to add, subtract, multiply, and divide whole numbers. You will also learn to use these computation skills in a variety of word problems.

Because word problems are important on math tests, this book pays special attention to the development of problem-solving skills. Your word-problem skills will increase as your computation skills increase.

Breakthroughs in Mathematics, Book 1 is divided into six chapters:

Chapter 1 is an introduction to the reading and writing of whole numbers.

Chapters 2 through 5 introduce the computation skills of addition, subtraction, multiplication, and division. While learning these skills, you will apply them in a variety of word problems.

Chapter 6, the last chapter in Book 1, is called "Special Topics In Mathematics." Here you'll study topics called "Introduction to Multi-Step Word Problems," "Introduction to Measurement," "Finding an Average," and "Squares, Cubes, and Square Roots." Because these topics are important in everyday life, they are included on many math tests.

Following this "To The Student" page is a three-page "Overview Of Skills." This overview will let you know which skills you are already familiar with and which you need to study carefully. Following each of Chapters 1 through 5 is a short "Review of Skills" that you can use to check your progress as you learn computation skills. Following Chapter 6 is a post-test to help you identify any skills in Book 1 with which you may need more practice.

On most pages, a row of "Skill Builders" comes before the problems. These Skill Builders are problems that have been started for you. Look at the work that has been done, then complete each problem.

To get the most out of your work, do each problem carefully. Check your answers to make sure you are working accurately. An answer key starts on page 169. Answers are given to all problems. In addition, the solutions to all word problems are worked out for you.

Overview of Skills

On the following three pages is an overview of the skills you will study in Book 1. This overview will help you evaluate your strengths and weaknesses in addition, subtraction, multiplication, division, and word problems.

How you do on this overview may help you decide where to spend most of your study time in this book. However, if you're preparing to take a math test in the future, it is a good idea to read the entire book.

Take your time as you work the problems. When you're finished, check your answers with the answers given on page 169.

Addition Skills

1. $\begin{array}{r} 4 \\ +3 \\ \hline \end{array}$	2. $\begin{array}{r} 9 \\ +0 \\ \hline \end{array}$	3. $\begin{array}{r} 4 \\ 1 \\ +2 \\ \hline \end{array}$	4. $\begin{array}{r} 43 \\ +21 \\ \hline \end{array}$
5. $\begin{array}{r} 23 \\ 12 \\ +\ 4 \\ \hline \end{array}$	6. $\begin{array}{r} 38 \\ +26 \\ \hline \end{array}$	7. $\begin{array}{r} 746 \\ 342 \\ +179 \\ \hline \end{array}$	8. $\begin{array}{r} 3,482 \\ 2,375 \\ +\ \ 358 \\ \hline \end{array}$

Subtraction Skills

9. $\begin{array}{r} 13 \\ -\ 8 \\ \hline \end{array}$	10. $\begin{array}{r} 53 \\ -21 \\ \hline \end{array}$	11. $\begin{array}{r} 86 \\ -19 \\ \hline \end{array}$	12. $\begin{array}{r} 482 \\ -\ 68 \\ \hline \end{array}$
13. $\begin{array}{r} 480 \\ -\ 29 \\ \hline \end{array}$	14. $\begin{array}{r} 307 \\ -174 \\ \hline \end{array}$	15. $\begin{array}{r} 900 \\ -269 \\ \hline \end{array}$	16. $\begin{array}{r} 1,060 \\ -\ \ 473 \\ \hline \end{array}$

Multiplication Skills

17. $\begin{array}{r} 7 \\ \times 4 \\ \hline \end{array}$	18. $\begin{array}{r} 32 \\ \times\ 3 \\ \hline \end{array}$	19. $\begin{array}{r} 44 \\ \times 12 \\ \hline \end{array}$	20. $\begin{array}{r} 43 \\ \times 20 \\ \hline \end{array}$
21. $\begin{array}{r} 730 \\ \times 302 \\ \hline \end{array}$	22. $\begin{array}{r} 78 \\ \times\ 6 \\ \hline \end{array}$	23. $\begin{array}{r} 76 \\ \times 49 \\ \hline \end{array}$	24. $\begin{array}{r} 385 \\ \times 193 \\ \hline \end{array}$

Division Skills

25. $8\overline{)48}$ **26.** $2\overline{)1,462}$ **27.** $4\overline{)80}$ **28.** $3\overline{)912}$

29. $5\overline{)125}$ **30.** $13\overline{)420}$ **31.** $31\overline{)6,479}$ **32.** $345\overline{)7,935}$

Word-Problem Skills

33. On Monday, Janice drove 174 miles. On Tuesday, she drove 250 miles. If she drove 195 miles more on Wednesday, how many total miles did she drive on these three days?

34. If Lucinda paid her grocery bill of $13.39 with a twenty-dollar bill, how much change should she receive?

35. At a sale price of $1.03 each, what does a total of twelve cans of orange juice concentrate cost?

36. If he is able to save $24 every month, how many months will it take Loren to save $336?

37. On her pay stub, Elsie noticed that her gross pay was $1,240. She also read that her deductions were as follows: federal tax, $61.00; state tax, $34.64; social security, $51.24; medical insurance, $10.00; and credit union, $45.50. What is the sum of her federal and state taxes?

38. On his diet, Thomas carefully measured all the food he ate. For dinner Tuesday, he cut a steak in two pieces. The uncut steak weighed 1 pound 7 ounces. If the piece he ate weighed 11 ounces, what is the weight of the piece he saved?

39. Each day, 7 days a week, Shauna practices 45 minutes of piano. Expressing your answer in hours and minutes, how much total time does Shauna practice each week?

40. What is the area of a garage floor that measures 7 yards long by 4 yards wide?

41. The carrying compartment on Peggy's moving van measures 6 feet high by 6 feet wide by 10 feet long. How many boxes, each 1 cubic foot in volume, can be packed into this cargo space?

42. On the three weekends in May that he played golf, Bryce had scores of 98, 111, and 106. What was Bryce's average score for these three rounds of golf?

Overview of Skills Evaluation Chart

On the chart below, circle the number of any problem you missed. The skill and study pages associated with each problem are indicated.

Problem Numbers	Associated Skills	Study Pages
ADDITION SKILLS		
1, 2, 3, 4, 5	Adding small numbers (no carrying)	18 to 23
6, 7, 8	Adding and carrying	32 to 36
SUBTRACTION SKILLS		
9, 10	Subtracting small numbers (no borrowing)	44 to 47
11, 12	Subtracting and borrowing	56 to 59
13, 14, 15, 16	Borrowing from zeros	60 to 63
MULTIPLICATION SKILLS		
17, 18, 19, 20, 21	Multiplying (no carrying)	72 to 83
22, 23, 24	Multiplying and carrying	84 to 90
DIVISION SKILLS		
25, 26, 27	Short division	100 to 104
28	Using zero as a place holder	107
29, 30, 31, 32	Long division	110 to 111

WORD-PROBLEM SKILLS

In addition to the page references given here, there are other pages in which word-problem skills are emphasized. If you miss any of the problems below, make sure you work carefully through all of the pages on word problems.

Problem Numbers	Associated Skills	Study Pages
33, 37	Addition word problems	30
34	Subtraction word problem	51
35	Multiplication word problem	82
36	Division word problem	125
38, 39	Measurement units word problems	} 140 to 159
40	Measurement: area word problem	
41	Measurement: volume word problem	
42	Averages word problem	160 to 161

1 Whole Numbers

Introduction to Whole Numbers

Since the beginning of human history, people have been fascinated with counting. Almost every modern culture has developed a way to count and a way to write counting symbols called **numbers**. Thousands of years ago, cave dwellers scratched patterns of marks on cave walls. These marks are the earliest known written numbers. Although these early numbers don't look at all like the numbers we use today, they served a similar counting purpose.

Centuries later, as societies became more educated, numbers became an important part of cultural life. Many early religious thinkers even came to believe that numbers had holy and magical powers.

Today we have a better understanding of numbers. We no longer think of them as holy or magical. However, we still realize their great importance in our own lives. Think for a moment of some numbers that may be of interest to you right now:

- the amount of money in your pocket

- the number of hours until dinner

- the income you'd like to earn next year

Each number stands for a certain amount or **value**. Given the choice, you'd probably prefer to earn $21,000 next year rather than $210. You'd choose the first because the symbol *21,000* stands for a much larger value than the symbol *210*.

The way we use symbols such as 21,000 and 210 to stand for different values is called our **number system**. You'll be studying our number system in this book. The main focus will be how numbers are added, subtracted, multiplied, and divided. Also, we'll discuss how numbers are used to solve word problems.

In this first chapter we'll briefly review the skills of **reading, writing, and rounding** whole numbers.

Digits and Place Value

Our number system uses symbols called **digits**. There are ten digits. From smallest to largest, the digits are **0, 1, 2, 3, 4, 5, 6, 7, 8,** and **9**.

Our counting numbers are called *whole numbers*. Whole numbers are formed by writing one or more digits in a row. The number 5 is a whole number. So are the numbers 210 and 21,000. 5 is a one-digit whole number, 210 is a three-digit whole number, and 21,000 is a five-digit whole number.

The *value of a digit* in a whole number depends on its place in that number. Because of this, digits are said to have **place value**. Place value increases as you start at the ones place at the right and move to the left. A 4 in the ones place has a smaller value than a 4 in the thousands place.

The first ten place values of whole numbers are shown below. Starting at the ones place and moving to the left, a comma is used to separate every group of three places.

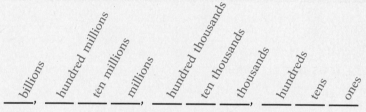

PLACE VALUES OF WHOLE NUMBERS

To find a digit's value, write the digit and follow it with the correct number of zeros to indicate its place value.

Example: What is the value of each digit in the number 4,207?

Step 1. 4 is in the thousands place.

4's value is **4 thousand**, which is written 4,000.

Step 2. 2 is in the hundreds place.
2's value is **2 hundred**, which is written 200.

Step 3. 0 is in the tens place.
However, 0's value is **0** in whatever place it appears.
Thus, there are no tens in the number 4,207.

Step 4. 7 is in the ones place.
7's value is **7 ones**, which is simply written 7.

What is the value of each underlined digit below? The first is done for you.

1. 4<u>8</u>5 *eighty*   **2.** 2,7<u>3</u>4 _____   **3.** 68<u>5</u> _____

4. <u>3</u>,385 _____   **5.** 7,8<u>9</u>0 _____   **6.** 23,<u>4</u>83 _____

7. <u>2</u>45,000 _____   **8.** 79,<u>0</u>00 _____   **9.** <u>3</u>,300,000 _____

Reading Whole Numbers

The value of a whole number is made up of the combined values of its digits. The examples below show how to read a whole number in this way.

Example 1: What is the value of 37?
    Step 1. 37 has **3** tens and **7** ones.
    Step 2. 37 is read "Thirty-seven."

Example 2: What is the value of 759?
    Step 1. 759 has **7** hundreds, **5** tens, and **9** ones.
    Step 2. 759 is read "Seven hundred fifty-nine."

Example 3: What is the value of 4,032?
    Step 1. 4,032 has **4** thousands, **0** hundreds, **3** tens, and **2** ones.
    Step 2. 4,032 is read "Four thousand, thirty-two."

Determine the number of thousands, hundreds, tens, and ones in each number below.

1. 18 has ____ tens and ____ ones.

2. 25 has ____ tens and ____ ones.

3. 36 has ____ tens and ____ ones.

4. 40 has ____ tens and ____ ones.

5. 137 has ____ hundreds, ____ tens, and ____ ones.

6. 782 has ____ hundreds, ____ tens, and ____ ones.

7. 309 has ____ hundreds, ____ tens, and ____ ones.

8. 720 has ____ hundreds, ____ tens, and ____ ones.

9. 4,982 has ____ thousands, ____ hundreds, ____ tens, and ____ ones.

10. 5,731 has ____ thousands, ____ hundreds, ____ tens, and ____ ones.

11. 8,093 has ____ thousands, ____ hundreds, ____ tens, and ____ ones.

12. 3,406 has ____ thousands, ____ hundreds, ____ tens, and ____ ones.

Determine the number of millions, hundred thousands, and ten thousands in each number below.

13. 5,420,000 has ____ millions, ____ hundred thousands, and ____ ten thousands.

14. 8,090,000 has ____ millions, ____ hundred thousands, and ____ ten thousands.

15. 4,700,000 has ____ millions, ____ hundred thousands, and ____ ten thousands.

Writing Whole Numbers

When writing whole numbers in words, there are three rules to remember:

- Numbers between twenty-one (21) and ninety-nine (99) are written with a hyphen (-) placed between the tens number and ones number.

  Examples: *Number in symbols*   *Number in words*

45	forty-five
27,000	twenty-seven thousand

- The word *and* is not used when writing whole numbers in words.

  Examples:

Number	Correct	Incorrect
126	one hundred twenty-six	one hundred and twenty-six
2,500	two thousand, five hundred	two thousand and five hundred

- A comma is placed after the word *million*, and after the word *thousand*, but *not* after the word *hundred*.

  Examples:

Number	Correct use of comma
12,471	twelve thousand, four hundred seventy-one
461,825	four hundred sixty-one thousand, eight hundred twenty-five
3,245,104	three million, two hundred forty-five thousand, one hundred four

  NOTE: Sometimes you will see a number over a thousand written without a comma. (For example, 5,352 may appear as 5362). Don't be confused by this. When *you* write a larger number, however, it is always a good idea to use a comma to separate the thousands from the hundreds.

Write each of the following numbers in words. Several are done for you.

1. 47 *forty-seven*

2. 39 _____

3. 56 _____

4. 92 _____

5. 238 _____

6. 579 _____

7. 735 _____

8. 830 *eight hundred thirty*

9. 3,587 _____

10. 5,865 _____

11. 7,329 _____

12. 23,254 *twenty-three thousand, two hundred fifty-four*

13. 38,675 _____

14. 73,184 _____

Writing Zero as a Place Holder

Zero is often used as a **place holder**:

- Placed in the middle of a number, zero shows that certain place values are not part of the number.

- Placed at the end of a number, one or more zeros make a great difference in a number's value.

Example 1: The number 309 contains 3 hundreds and 9 ones.
The 0 shows that there are no tens.

Example 2: The amounts $210 and $21,000 differ only by two zeros. But look at the difference in value! The 2 in $210 stands for 2 hundred, while the 2 in $21,000 stands for 2 ten thousands.

When writing a number in symbols, write a zero in each place that is not expressed in words as part of the number. Look at the following examples.

Examples:

Number in words	Number in symbols
two hundred seven (0 tens)	207
seven thousand, sixty-five (0 hundreds)	7,065
one hundred twenty-three thousand, nine (0 hundreds and 0 tens)	123,009
two million, nineteen thousand, six hundred (0 hundred thousands, 0 tens, and 0 ones)	2,019,600

Write each of the following number phrases in symbols. Remember to place a comma after millions and thousands.

1. three hundred eight _____

2. five hundred seven _____

3. three thousand, fifteen _____

4. forty-six thousand, eleven _____

5. two hundred nine thousand, four hundred _____

6. six hundred four thousand, two hundred _____

7. eight hundred seventeen thousand, six hundred five _____

8. six million, three hundred seven thousand _____

9. eight million, two hundred one thousand _____

10. nine million, seven hundred two thousand, thirteen _____

Rounding Whole Numbers

Often, it is useful to **approximate**. To approximate is to give a number that is "about equal" to an accurate amount. Here is an example:

- A football sports announcer says that 75,000 people are attending a football game. (Actually, 74,839 people are there.)

The number 75,000 is called a **_round number_**. A round number has zeros to the right of a chosen place value. The number 75,000 has zeros to the right of the thousands place. Thus, 74,839 rounded to the thousands place is 75,000.

One way to round a number is to identify two round numbers that the exact number lies between. Then you simply ask, "Which of these two round numbers is the exact number closest to?"

Example 1:   Round 74,839 to the nearest 100.

Step 1.   To the hundreds place, 74,839   **74,800** ← 74,839 → 74,900
lies between 74,800 and 74,900.
                                                 74,839 is less than halfway
Step 2.   Since 74,839 is closer to 74,800,   between 74,800 and 74,900.
74,800 is the answer.

Another way to round a number is to follow the steps given below:

STEPS FOR ROUNDING WHOLE NUMBERS

1. Underline the digit in the place you are rounding to.
2. Look at the digit to the right of the underlined digit.
   a) If the digit to the right is 5 or more, add 1 to the underlined digit.
   b) If the digit to the right is less than 5, leave the underlined digit as it is.
3. Put zeros in all places to the right of the underlined digit.

Example 2:   Round $3,753 to the nearest hundred dollars.

Step 1.   Underline the digit in the hundreds place.

underline the 7: 3,7̲53

Step 2.   Look at the digit to the right of the 7.
The digit is 5. Since 5 is equal to "5 or
more," add 1 to the underlined digit 7.

add 1 to 7: 3,8̲5̶3̶

Step 3.   Now put zeros in all places to the right of 8.

replace 53 with 00: 3,800

Answer: **$3,800**

Round each number below as indicated. For each number, circle one of the two answer choices.

To the nearest 10

1. 53: 50 *or* 60   86: 80 *or* 90   75: 70 *or* 80   124: 120 *or* 130

To the nearest 100

2. 239: 200 *or* 300   462: 400 *or* 500   2,475: 2,400 *or* 2,500

To the nearest 1,000

3. 4,627: 4,000 *or* 5,000   5,820: 5,000 *or* 6,000   13,381: 13,000 *or* 14,000

Using either method, round the following numbers.

To the nearest 10

4.   34 _____   82 _____   125 _____ $281 _____

To the nearest 100

5.   239 _____ $391 _____   4,712 _____ $5,351 _____

To the nearest 1,000

6. 4,851 _____   5,384 _____ $8,901 _____ $12,499 _____

7. Gina earns $1,287 each month. Round Gina's salary to the nearest hundred dollars.

8. Maximum seating at Dino's Pizza is 72 people. Round this number to the nearest 10 people.

9. The average distance from the earth to the sun is 92,900,000 miles. Round this distance to the nearest million miles.

10. The average cost of a three-bedroom house in New Castle is $79,450. Round this cost to the nearest thousand dollars.

Working with Dollars and Cents

Part of our study of whole numbers includes working with money. Here we'll briefly discuss the ways we read and write dollars and cents.

There are three common ways to write dollars and cents:

Words only	Number/word-label	Symbols only
five dollars	5 dollars	$5
twelve dollars	12 dollars	$12
eighty cents	80 cents	80¢
nine cents	9 cents	9¢

Most often we work with money that combines dollars and cents. To write the total, we use a dollar sign and a ***decimal point***. The decimal point separates the number of dollars from the number of cents. In decimal form, twelve dollars and thirty-five cents is written as

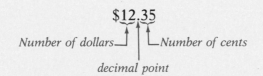

$$\$12.35$$

Number of dollars⌐ ⌐Number of cents

decimal point

United States Money
COINS

50¢/$0.50   25¢/$0.25

10¢/$0.10 5¢/$0.05 1¢/$0.01
COMMON BILLS

$50 $20

$10 $5

$2 $1

There are five facts to remember about writing dollars and cents in decimal form:

1. Dollars are written to the left of the decimal point.
2. Cents are written in the first two places to the right of the decimal point.
3. When reading (or writing) dollars and cents in decimal form, you treat the decimal point as the word *and*.
4. When the amount of cents is between 1¢ and 9¢, write a zero between the decimal point and the cents digit.
   <u>Example</u>: Four dollars and six cents is written $4.06.
5. When writing only cents in decimal form, you can write it with or without a zero in the dollar column.
   <u>Example</u>: Nineteen cents can be written $0.19 or $.19.

Write each of the following amounts using a dollar sign and decimal point. The first amount is completed as an example.

1. Twenty dollars and eight cents   *$20.08*
2. Thirty-two dollars and nine cents   _____
3. Sixty-one dollars and fifteen cents   _____
4. Eighteen dollars and thirty-eight cents   _____
5. Fifty-two dollars and three cents   _____
6. Twenty-five dollars and seventy-five cents   _____
7. One hundred fifteen dollars and fifty cents   _____
8. Seventy-nine dollars and eighty cents   _____
9. Three hundred nine dollars and four cents   _____
10. Eighty-seven dollars and fifty-five cents   _____

Write each amount below. On the first line, use the cents sign, and on the second line, use the dollar sign and decimal point.

11. eight cents   *8¢*   *$.08*
12. nine cents   _____   _____
13. four cents   _____   _____
14. three cents   _____   _____
15. thirty-five cents   _____   _____
16. forty-three cents   _____   _____
17. ninety-eight cents   _____   _____
18. fifty-seven cents   _____   _____

On each line below, fill in the amount shown, using words only. The first one is done for you.

19. $5.07   *Five dollars and seven cents*
20. $12.06   _____
21. $7.37   _____
22. $23.65   _____
23. $125.50   _____

Rounding Dollars and Cents

You round dollars and cents in the same way you round whole numbers. You can round a monetary amount to the nearest ten cents (dime), nearest dollar, nearest ten dollars, and so on.

The two places to the right of the decimal point represent <u>cents.</u>
The places to the left of the decimal point represent <u>dollars.</u>

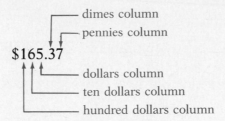

Example: Round $15.76 to the nearest dime.

Step 1. Underline the digit in the dimes column.

underline the 7: $15.7<u>6</u>

Step 2. Look at the digit to the right of the 7. The digit is 6. Since 6 is more than 5, add 1 to the underlined digit 7.

add 1 to 7: $15.8̶6̶

Step 3. Now put zeros in all places to the right of 8.

replace 6 with 0: $15.80

Answer: $15.80

Round each of these amounts to the nearest dime.

1. $5.79 _____ $1.63 _____ $8.08 _____ $10.34 _____

Round each of these amounts to the nearest dollar.

2. $6.72 _____ $13.49 _____ $52.91 _____ $9.08 _____

In each problem below, round each number as indicated.

3. A bottle of pop costs $0.89. Round this amount to the nearest 10¢.

4. Disposable razors went on sale for 74¢ per package. Round this amount to the nearest dime.

Reading, Writing, and Rounding Skills Review

The problems below will give you a chance to briefly review the skills of this chapter. Work each problem as carefully as you can and check your answers with those given on page 170. Correct any mistakes.

UNDERSTANDING PLACE VALUE: Review page 8.

What is the value of each underlined digit below?

1. 5<u>9</u>8 _____   <u>2</u>,397 _____   38<u>6</u> _____   4,<u>5</u>90 _____

Determine the number of thousands, hundreds, tens, and ones in each number below.

2. 5,038 has ____ thousands, ____ hundreds, ____ tens, and ____ ones.

3.   927 has ____ thousands, ____ hundreds, ____ tens, and ____ ones.

WRITING WHOLE NUMBERS: Review pages 10 and 11.

Write each of the following numbers in words.

4.   163 _____

5. 4,079 _____

6. 27,000 _____

Write each of the following number phrases as numbers.

7. five hundred seventy-five   _____

8. two thousand, forty-six   _____

9. four million, six hundred thousand   _____

ROUNDING WHOLE NUMBERS AND DOLLARS AND CENTS: Review pages 12 through 16.

Round each of the following numbers to the place indicated.

10. 76 (tens place)   11. 164 (hundreds place)   12. 2,590 (thousands place)

13. $0.37 (nearest dime)   14. $3.48 (nearest dollar) 15. $67.09 (nearest ten dollars)

2 Addition Skills

Concepts in Addition

To add is to put two or more numbers together. The concept of adding is already familiar to you. Here are two examples:

- You add the value of coins in your pocket to see how much change you have.

- You add the prices of items in your grocery cart to figure out how much you'll have to pay.

To begin this chapter, we want to point out five other concepts that will make your study of addition more meaningful.

1. In symbols we represent addition by the plus sign (+). For example, the words *five plus three* can be written as 5 + 3.
   You can write an addition problem in a column (up and down) or in a row (across). Numbers being added are called **addends**. The answer is called the **sum** or **total**.

   Written in a column

   $$\begin{array}{r} 5 \leftarrow \text{addend} \\ + \ 3 \leftarrow \text{addend} \\ \hline \leftarrow \text{sum or total} \end{array}$$

   Written in a row

   5 + 3 = 8

   addends   sum or total

2. To add means to put <u>like</u> things together. Like things are of the same type. We can add two or more amounts of money; we can add two or more groups of cars. But we can't add cars to money! Adding unlike things doesn't make sense.

   You can add like things.   You cannot add unlike things.

Adding with Empty Columns

An empty place in a column is treated as a zero. An empty digit place means there is nothing to add.

In this example, only a single digit is in the hundreds column. There is nothing to add to it from the bottom number. Thus, 4 is written in the hundreds place in the answer.

Example: Add:

$$\begin{array}{r} 4\,35 \\ +\,60 \\ \hline 4\,95 \end{array}$$

Step 1. 5 + 0 = 5

Step 2. 3 + 6 = 9

Step 3. 4 + 0 = 4

Add. Treat each empty digit place as a zero.

1.

13	41	20	251	342
+ 6	+ 5	+ 8	+ 36	+ 55

2.

306	734	392	282	425
+ 192	+ 45	+ 104	+ 205	+ 61

3.

3,452	4,703	5,611	7,307	4,256
+ 304	+ 1,095	+ 288	+ 481	+ 1,402

4.

7,005	21,450	30,536	2,125	26,300
+ 1,970	+ 6,237	+ 8,342	+ 1,431	+ 3,657

5.

32	13	231	112	540
13	2	17	101	214
+ 4	+ 4	+ 11	+ 23	+ 21

6.

2,415	3,250	6,023	11,735	14,500
132	1,112	1,445	2,223	12,156
+ 201	+ 400	+ 2,211	+ 1,021	+ 2,031

Adding Dollars and Cents

To add dollars and cents, line up the decimal points and then add the columns as you did when adding whole numbers. Line up the decimal points by placing one directly below the other.

Example 1: Add $2.56 and $1.32.

    Step 1. Write the numbers in columns. Line up the decimal points.

    Step 2. From right to left, add the columns.
        Pennies column: 6 + 2 = 8
        Dimes column: 5 + 3 = 8
        Dollars column: 2 + 1 = 3

```
                dollars column
                 dimes column
                  pennies column
            $2.56
             1.32
            $3.88
              Decimal points
              are lined up.
```

Answer: **$3.88**

Example 2: Add 65¢ and 2¢.

```
   65¢          $0.65
 +  2¢   OR   +  0.02
   67¢          $0.67
```

Answer: **67¢** OR **$0.67**

Example 3: Add $1.14 and 82¢.

    Step 1. Write 82¢ as a decimal.
        82¢ = $0.82

    Step 2. Add:
```
        $1.14
      +  0.82
        $1.96
```

Answer: **$1.96**

NOTE: In Example 2, the answer can be expressed either in cents or in decimal form. In Example 3, the 82¢ must first be written in decimal form before the addition can be done.

Add. Remember to include the decimal point and dollar sign in each answer.

1.
```
  $5.23      $1.38      $6.50      $4.75      $2.15      $9.08
+  2.35    +  3.20    +  2.25    +  3.10    +  0.44    +  0.71
```

2.
```
  $12.52     $32.09     $41.65     $24.45     $15.80     $25.25
+   2.31   +  12.50   +  14.32   +   5.13   +  20.17   +  10.00
```

Add. Write each answer two ways. First use the cents symbol and then use a dollar sign and decimal point.

3. Add 25¢ and 14¢   Add $0.63 and $0.16   Add 73¢ and 25¢

Add. As your first step, express each number in decimal form.

·4. $4.52 + 43¢ $1.25 + 31¢   75¢ + $1.10

    $2.43 + 46¢   81¢ + $3.14

Write the sum represented by each group of coins below. First write the sum as a number of cents using the cents symbol. Then write the sum in decimal form. The first problem is completed as an example.

5.

    _76¢_   _$0.76_   _____   _____

6.

    _____   _____   _____

Using a dollar sign and decimal point, write each sum of money below.

7.

    _____   _____   _____

8.

    _____   _____   _____

27

Introducing Word Problems

What Is a Word Problem?

A **word problem** is a short story that asks a question or tells you to find something. You are given information and are asked to set up and solve a problem.

Here is an example of a word problem:

During March it rained 7 inches. April and May each had 5 inches of rain. What was the total rainfall during these three months?

Understanding the Question

Word problems may be written as a single sentence, or as several sentences. In each type of problem, *the first step is to identify the question and to understand what you are being asked to find.*

Example 1 is a single-sentence problem. The sentence contains numbers and asks a question.

Example 1:   *How much does it cost to bowl 3 lines if each line costs $1.15?*

asked to find: cost of bowling 3 lines

Example 2 is a multi-sentence problem. The last sentence tells you what you are being asked to find, while other sentences contain numerical information.

Example 2:   Janice jogs five days a week. On Monday, Wednesday, and Friday, she takes 3-mile runs. On Tuesday and Thursday, she takes 5-mile runs. *Find how many miles Janice jogs each week.*

asked to find: miles jogged each week

In Example 3 you must *interpret* (figure out the meaning of) what the question asks you to find.

Example 3:   In Donna's kindergarten class, there are 13 girls and 9 boys. *How many kids are in this class?*

asked to find: number of kids in class

Although the word *kids* appears only in the question, you interpret it to mean *both* boys and girls.

Underline the question in each problem below. Then circle the words within the parentheses that identify what you are asked to find. DO NOT SOLVE THESE PROBLEMS.

1. If Sarah pays $175 every two weeks for rent, how much rent does she pay each month?

   (weekly rent, monthly rent, yearly rent)

2. How much do two gallons of milk weigh if one weighs 9 pounds?

   (number of gallons, weight of 9 gallons, weight of 2 gallons)

3. This year, Manuel paid $850 in federal income tax, $426 in state income tax, and $47 in city income tax. Find the amount that Manuel paid in state and federal taxes.

   (federal and state taxes, city tax, total taxes)

4. Joan bought the following items at the store: 1 can of pop for 89¢, 1 gallon of milk for $1.92, and 2 packages of gum for 50¢. How many items did Joan buy?

   (number of items, cost of groceries, cost of food items)

5. Road distance between Oak Grove and Rockville is 132 miles. From Rockville to Salem is 91 miles. Driving through Rockville, how many miles is it between Oak Grove and Salem?

   (miles from Rockville to Salem, miles from Oak Grove to Salem, miles from Oak Grove to Rockville)

In the following problems, circle the words within the parentheses that are *closest in meaning* to what you are asked to find.

6. If 9 cats and 14 dogs were in the First Grade Pet Parade, how many pets did the kids bring?

   (number of cats, number of dogs, number of cats and dogs, number of kids)

7. At a variety sale, Stacey bought a blouse for $9, a skirt for $17, a plant for $6, and a vase for $5. Find how much Stacey paid for the clothes she bought.

   (cost of all items, cost of skirt and blouse, number of items)

8. During the week, Jason worked a 35-hour shift, earning $9 per hour. He also worked 3 hours of overtime, earning $13 an hour. Find how much extra money Jason earned this week.

   (total pay, total regular pay, total overtime pay)

Addition Word Problems

On this page and the next, there are addition problems for you to solve. In each problem, pay close attention to what the problem asks you to find.

To write an answer correctly, remember to include both a number and a label. The label tells what the number refers to. A label may be a word such as *pounds* or *years*, or it may be a symbol such as $ or ¢.

Example: A shirt is on sale for $4.25, and a pair of socks is on sale for $1.50. If Bill bought both, how much did he spend?

asked to find: total amount spent

Step 1. Add the numbers:
$$\begin{array}{r} 4.25 \\ +1.50 \\ \hline 5.75 \end{array}$$
Check:
$$\begin{array}{r} 1.50 \\ +4.25 \\ \hline 5.75 \end{array} ✔$$

Step 2. Attach the label $.

Answer: **$5.75**

In each addition problem below, underline the words in the question that identify what you are asked to find. Then solve and check each problem. Problem 1 is completed as an example.

1. If a class contains 23 men and 16 women, <u>how many students are in the class?</u>

   $$\begin{array}{r} 23 \text{ men} \\ +16 \text{ women} \\ \hline 39 \text{ students} \end{array} \qquad \begin{array}{r} 16 \\ +23 \\ \hline 39 \end{array}✓$$

2. The distance from Seattle to Portland is 172 miles. If Portland is 114 miles from Eugene, how far is Seattle from Eugene? (Hint: Put the mileage on the diagram below.)

   Seattle   Portland   Eugene

3. At a price of 23¢ each, find the cost of three candy bars.

4. In his toolbox, George has 4 Phillips screwdrivers and 5 slothead screwdrivers. How many screwdrivers does he have altogether?

5. David paid $1,245 more for his car than he sold it for. If he sold it for $6,250, what was David's purchase price?

6. In Tina's pocket are several coins: one quarter, three dimes, and four pennies. Find how much money Tina has in change.

7. If it is now 12 minutes after 8 o'clock, what time will it be in 27 minutes?

8. Ira Thompson makes a monthly rent payment of $415. If his other monthly expenses are $581, what take-home monthly income does he need to pay these expenses?

9. At 9:00 A.M., the temperature was 43°F. By noon it had warmed up 14 degrees more. What was the temperature at noon?

10. How much did Mrs. Murphy pay for two hats if she bought a hat for herself for $8.25, and one for her daughter for $9.50?

11. On Barbara's 13th birthday, she weighed 101 pounds. During the next two years, she gained 14 more pounds. How much did Barbara weigh on her 15th birthday?

12. Evening phone rates in Erie are discounted. If the cost of the first minute is 30¢, and the cost of each additional minute is 21¢, what is the cost of a three-minute call?

13. If it rained 3 inches on Monday, 5 inches on Tuesday, and 4 inches on Wednesday, how many inches of rain fell on the first three days of the week?

14. What was the total of John's three bills if his rent was $405; telephone, $21.14; and electricity, $71.25.

15. For their first basketball game of the season, the Bulldogs drew an attendance of 2,324 people. The second game drew 3,042 people, and the third game drew 3,121. Find the combined attendance for these first three games.

Adding and Carrying

Adding two-digit or larger numbers often involves **carrying**. To *carry* means to take a digit from the sum of one column and place it at the top of the column to the left. The example below shows how to carry 10 ones as 1 ten.

Example: Add: 45   Solution: $\overset{1}{4}5$ Check: $\overset{1}{3}9$
                      +39   +39   +45
                                84   84 ✔

Step 1. Add the ones column: $5 + 9 = 14$
Think of the 14 ones as 1 ten and 4 ones.
Place the 4 under the ones column.

Step 2. Carry the 1 ten to the top of the tens column.

Step 3. Add the tens column: $1 + 4 + 3 = 8$
Place the 8 under the tens column.

┌─ Carry 1 ten to top
│   of tens column
↓
$\overset{1}{4}5$
+39
84
└─ Place 4 under
   ones column.

Add each row of problems below. Check some of your answers on scratch paper to make sure you are adding and carrying correctly.

Carrying to the Tens Column

Skill Builders

1.

$\overset{1}{2}7$	$\overset{1}{4}5$	$\overset{1}{6}4$	$\overset{1}{7}2$	$\overset{1}{3}6$	$\overset{1}{2}4$
+ 6	+ 9	+28	+59	23	13
				+17	+45
3	4	2	1	6	2

2.

48	63	52	69	77	35
+ 7	+ 9	+ 8	+36	+48	+29

3.

$35	27¢	$42	$78	65¢	52¢
+ 7	+ 8¢	+ 9	+ 15	+16¢	+18¢

4.

14	31	27	25	38	56
5	9	16	14	26	47
+ 3	+ 7	+ 6	+16	+19	+29

Carrying to the Hundreds Column

When the sum of digits in the tens column is 10 or larger, carry to the hundreds column.

Example:

```
  ¹
  2 9 5
+   7 3
  3 6 8
```

Skill Builders

5.

				¹ 274	² 990
¹ 325	¹ 564	725	746	122	261
+ 84	+ 71	+192	+482	+ 33	+183
09	35	7	8	29	34

6.

274	883	497	273	928	742
+ 62	+ 95	+ 31	+236	+590	+385

7.

436	271	134	534	556	965
251	162	180	420	482	871
+ 72	+ 34	+103	+191	+181	+590

Carrying to the Thousands Column

When the sum of digits in the hundreds column is 10 or larger, carry to the thousands column.

Example:

```
    ¹
  6, 7 34
+ 2, 8 41
  9, 5 75
```

Skill Builders

8.

				² 6,724	4600
¹ 8,263	¹ 4572	5,663	8933	1,931	3715
+ 920	+ 725	+2,821	+2731	+ 623	+2920
183	297	84	64	278	35

9.

4561	5732	2754	6362	8664	8451
+ 925	+ 656	+1825	+2935	+4602	+7843

10.

5,340	2,632	2,513	4,435	5,736	8,660
2,125	1,221	1,243	3,520	2,920	7,815
+ 803	+ 225	+1,620	+3,414	+1,812	+6,903

Carrying to the Tens <u>and</u> Hundreds Columns

The following example shows how to carry to *both* the tens and hundreds columns.

Example: Add:
$$
\begin{array}{r}
\overset{1\ 2}{4}38 \\
247 \\
+\ 79 \\
\hline
764
\end{array}
$$
Check:
$$
\begin{array}{r}
\overset{1\ 2}{}79 \\
247 \\
+438 \\
\hline
764\ \checkmark
\end{array}
$$

Ones column: Add the ones column: $8 + 7 + 9 = 24$
Place the 4 under the ones column.
Carry the 2 to the tens column.

Tens column: Add the tens column:
$2 + 3 + 4 + 7 = 16$
Place the 6 under the tens column.
Carry the 1 to the hundreds column.

Hundreds column: Add the hundreds column:
$1 + 4 + 2 = 7$
Place the 7 under the hundreds column.

Skill Builders

11.

$\overset{1\ 1}{5}82$	$\overset{1}{3}57$	$\overset{1\ 1}{7}93$	$\overset{1}{7}38$	$\overset{1\ 1}{4}45$	$\overset{1}{7}84$
+ 39	+ 65	+137	+577	352	546
				+156	+ 97
21	2	30	5	53	7

12.

674	458	117	573	263	489
+ 87	+ 46	+ 85	+ 89	+ 49	+ 37

13.

576	355	495	674	762	587
+265	+249	+375	+276	+158	+384

14.

858	765	379	637	759	946
+475	+737	+894	+563	+748	+164

15.

285	357	187	830	787	896
123	130	156	707	473	839
+ 47	+ 98	+ 79	+283	+365	+780

In many addition problems, you may need to carry to several columns in the same problem. The problems below give you practice in developing this skill.

Carrying to Several Columns

16.

			¹¹¹ 4712	¹ 5,385
¹ 3,647	¹¹ 7825	¹ 6,764	3388	2,482
+ 691	+1776	+5,874	+ 825	+1,300
338	01	38	925	67

17.

4,657	5,624	8,351	3,573	5,374
+ 572	+ 895	+ 769	+ 589	+ 887

18.

5457	5284	7572	6571	8582
+2872	+3947	+1679	+3788	+3248

19.

1,572	5,682	3,902	7,478	6,684
945	875	2,475	3,374	5,985
+ 570	+ 785	+ 860	+2,362	+3,481

20.

			¹¹¹ 75400	¹¹¹ 45,455
¹¹ 12,745	¹¹¹ 14375	¹¹ 53,635	46471	36,275
+ 8,480	+ 7936	+19,565	+13836	+23,450
225	311	00	5707	180

21.

31284	54264	26473	42382	45690
+ 9785	+ 8890	+ 7978	+14899	+26750

22.

10,357	15,683	42,497	56,386	67,389
6,782	8,435	21,470	34,150	52,500
+ 3,800	+ 4,567	+ 9,233	+10,854	+24,687

Adding Numbers in a Row

When numbers are written in a row, rewrite them in a column before adding. Most people find it easiest to place the largest number at the top of the column and smaller numbers at the bottom. Be sure to place ones digits in a ones column, tens digits in a tens column, and so on.

Example: Add: 452 + 78 + 326

Write the numbers in a column. Line up the hundreds digits (4 and 3), tens digits (5, 2, and 7), and ones digits (2, 6, and 8) directly over one another. Add.

$$\begin{array}{r} \overset{\scriptstyle 1\ 1}{452} \\ 326 \\ +\ \ 78 \\ \hline 856 \end{array}$$

Rewrite each of the following problems in a column and add.

1. 28 + 9   17 + 6   15 + 13

2. 36 + 8 + 24   42 + 27 + 14   25 + 19 + 33

3. 37 + 156 + 40   163 + 57 + 38   75 + 428 + 126

4. 465 + 97 + 873   783 + 588 + 180   2,340 + 746 + 983

5. 5,354 + 3,365 + 984   23,837 + 7,486 + 5,386

Carrying with Dollars and Cents

Carrying with dollars and cents is done the same way as carrying with whole numbers. You carry a digit across the decimal point as if it weren't there. Remember to place a decimal point and dollar sign in the answer.

Example: Add: $5.75   Solution: $\overset{1}{\$}5.75$ Check: $\overset{1}{\$}2.81$

                   + 2.81   + 2.81   + 5.75

                                     $8.56 $8.56 ✔

Pennies column: Add the pennies column: 5 + 1 = 6
                       Place the 6 under the pennies column.

Dimes column: Add the dimes column: 7 + 8 = 15
                       Place the 5 under the dimes column.
                       Carry the 1 to the dollars column.

Dollars column: Add the dollars column: 1 + 5 + 2 = 8
                       Place the 8 under the dollars column.

Bring the decimal point straight down in the sum and add a dollar sign.

Add as indicated.

1. $8.95 $9.77 $14.37 $24.58 $30.75 $12.48
     + 1.57   + 3.89   + 3.74   + 18.63   + 9.85   + 9.75

2. $2.37 + 95¢ $3.85 + 87¢ $10.50 + 99¢ $14.75 + 79¢

Add the two sums of money in each problem below.

3.

4.

37

Finding Necessary Information

The first step in solving a word problem is understanding what the question asks you to find. *The second step is deciding what information is needed to answer the question.*

What Is Necessary Information?

In real-life situations (and on tests!), you may see word problems that contain many numbers and labels. Often you are given more information than is needed to answer the question. Therefore, it is important that you be able to select only that information which you need.

Necessary information includes *only* those numbers and labels needed to answer a specific question.

Given information includes *all* of the numbers and labels that appear in a problem.

Necessary information may appear anywhere in a problem. In order to solve the problem, ask yourself, "What information do I need to answer this question?" Then reread the problem carefully and select those numbers that are needed.

Example 1: What is the combined weight of Bill who weighs 87 pounds and his twin brother Bob who weighs 91 pounds?

given information: 87 pounds, 91 pounds

necessary information: 87 pounds, 91 pounds

In Example 1, you need to use all the information given. Both weights are needed to answer the question, and no other information is given.

Example 2: Lucy planned to spend no more than $20 at the grocery store. She bought a roast for $7.85, vegetables for $5.60, and milk for $1.95. How much did Lucy spend in all?

given information: $20, $7.85, $5.60, $1.95

necessary information: $7.85, $5.60, $1.95

In Example 2, the question asks only about how much Lucy spent. Lucy's spending limit of $20 is not part of the grocery bill. Thus, $20 is **extra information**, information not needed to solve the problem.

In each problem below, underline what you are being asked to find. Then write *only* the information needed to solve the problem on the line below. Be sure to include labels. DO NOT SOLVE THESE PROBLEMS. The first one has been done for you.

1. <u>What is the total price of a quart of milk</u> that costs $0.48 <u>and a loaf of bread</u> that costs $1.09?

    <u> $0.48 $1.09   </u>

2. While shopping Tuesday, Juanita bought a dress for $34.95, a skirt for $19, a picture frame for $15.75, and a glass bowl for $9.95. What amount did she spend for the clothes?

    _____

3. Brenda has lost 13 pounds on her diet. Her weight dropped from 167 pounds to 154 pounds. If she still wants to lose 14 more pounds, how much does Brenda want to lose in all?

    _____

4. How many total hours does Amy work each week if she earns $278 per week working 35 hours as a cook and $2 per hour baby-sitting 8 hours each week?

    _____

5. What was the total rainfall during the first 3 days of the week if during the week it rained every day? Monday had 2 inches of rain; Tuesday, 1 inch; Wednesday, 2 inches; Thursday, 3 inches; and Friday, 2 inches.

    _____

6. Brad bowled 3 games Tuesday night. His scores were 147, 172, and 165. These scores were well below his league average of 192. Find the sum of his Tuesday night scores.

    _____

7. In Canyonville there are 23,458 registered voters. Of these, 13,228 are men and 10,230 are women. During the November election, 4,572 men voted. If 6,385 women also voted, how many people voted in this election?

    _____

8. What is the total cost of a 7-lb. package of chicken when chicken is on sale for $.89 a pound and hamburger is on sale for $1.69?

    _____

Addition Word Problems

On these next two pages are problems that give you practice applying addition skills. In each problem, watch for words that let you know you should add to find the answer. Examples are words such as *total, altogether,* and *sum.* As you begin each problem, remember to . . .

- read each problem carefully

- understand what the question is asking you to find

- select *only* the necessary information

Solve each addition problem below.

1. In a 50-mile bike race, Sally rode 12 miles the first hour, 9 miles the second hour, and 13 miles the third hour. How many miles altogether did Sally ride during these first 3 hours?

2. Tony lost 9 pounds during the first two months of his diet. He would still like to lose 11 more pounds and see his weight drop to 165 pounds. How many total pounds does Tony hope to lose?

3. Wilma paid $6.95 for a 10-pound bag of fertilizer for her lawn. When that ran out, she bought a 25-pound bag for $14.95. What was the total amount paid for fertilizer?

4. Working as a waitress, Julie earns $36 a day in salary and about $18 a day in tips. Tuesday was a very good day, and Julie earned $24.75 in tips. What total earnings did Julie take home Tuesday?

5. There are 14,842 Democrats and 17,463 Republicans registered in Cottage Grove. In the March election, 6,284 Democrats and 5,782 Republicans voted. Only 800 Independents voted. Find how many people altogether voted in this election.

6. On a trip, Nancy kept a record of her gas use. Her record included: 14 gallons — 209 miles; 17 gallons — 248 miles; and 9 gallons — 173 miles. How many total miles did she drive on her trip?

7. Registration for a cooking class cost Thomas $21. He also needed a book. A new book cost $18.95, while a used book cost $8.75. If he bought the used book, how much would Thomas spend for the class altogether?

8. How many cents out of each dollar do the Johnsons spend for the sum of rent, car, and food expenses? For each dollar the Johnson family earns, they spend 30¢ for rent, 11¢ for medical care, 13¢ for car expense, 21¢ for food, 9¢ for clothes, 8¢ for recreation, and 7¢ for savings and miscellaneous.

9. For a barbecue party, Fred bought 3 pounds of chicken, 4 pounds of hamburger, 5 pounds of steak, and 8 pounds of potato salad. How many pounds of meat did Fred buy altogether if he planned to feed between 30 and 40 people?

10. While shopping, Cal bought car oil for $11.95, a shovel for $9.49, dish soap for $1.79, and a rake for $13.50. Find how much Cal paid for the two garden tools.

11. Mary is now 14 years old. Her sister, Lila, is 23 and just had a baby girl, Sandra. How old will Mary be when Sandra starts school at age 6?

12. Jimmy earns $75.50 for each used car he sells. On Tuesday, he sold 2 cars, 1 for $900 and 1 for $650. On Friday, he sold 1 for $1,250. What total amount did Jimmy earn for these sales?

13. For lunch, Rose had a ham sandwich for $2.49, a glass of milk for 65¢, and ice cream for 75¢. After seeing her bill, which included a tax of $0.21, she decided to pay with a $5 bill. What was the total cost of Rose's lunch?

14. Frank looked at 3 used cars: a Plymouth for $1695, a Chevrolet for $1545, and a Ford for $1275. He chose the Plymouth after they lowered the price to $1475. He then put on new tires for $276 and put in a radio for $125. What total amount did Frank invest in this car?

Addition Skills Review

Before leaving the chapter on addition, complete these next two pages to briefly review your addition skills. Work each problem as carefully as you can. Check your answers with those given on pages 172–173. Correct any mistakes.

ADDING SINGLE DIGITS: Review pages 20 through 23.

1.
$$6 + 2 \qquad 8 + 7 \qquad \begin{array}{r} 9 \\ 6 \\ +4 \end{array} \qquad 5 + 7 = \qquad 8 + 9 + 2 =$$

ADDING LARGER NUMBERS, NO CARRYING: Review pages 24 through 27.

2.

23 +16	78 +21	143 + 52	634 121 + 23	400 172 +106	3,480 1,002 + 315

ADDING AND CARRYING: Review pages 32 through 35.

3.

27 + 6	43 + 9	205 + 58	457 + 86	687 +159	2,380 + 638

4.

75 48 + 8	87 36 +21	106 97 + 82	500 396 +208	1,339 835 + 400	13,298 9,509 + 4,582

ADDING DOLLARS AND CENTS: Review pages 26–27 and page 37.

5.

$4.23 + 1.10	$6.13 + .72	$7.46 + 2.39	$145.79 + 25.89	$2,345.65 1,394.05 + 375.37

ADDING NUMBERS WRITTEN IN A ROW: Review pages 23 and 36.

6. 16 + 8   47 + 9   35 + 28   754 + 395

7. 32 + 18 + 5   54 + 30 + 17   343 + 105 + 84

RECOGNIZING ADDITION WORD PHRASES: Review pages 19, 28, and 29.

Write each word phrase in symbols and then solve.

8. fourteen plus nine   **9.** the sum of twenty-five dollars and six dollars   **10.** thirty-eight pounds added to sixty-seven pounds

Write each addend in decimal form and then solve.

11. $0.45 plus 37¢   **12.** the sum of $2.34 and 96¢   **13.** 84¢ plus 59¢

SOLVING WORD PROBLEMS: Review pages 28 through 31 and pages 38 through 41.

14. What is the sum of 2 quarters, 3 dimes, and 7 pennies?

15. While moving furniture, Ramos moved 9 chairs, 7 tables, and 3 stools. How many pieces of furniture did Ramos move altogether?

16. John bought a wrench for $9.89, a pair of pliers for $4.95, and a drill bit for $2.39. How much did John spend in all for these tools?

17. Walton Stadium holds 46,000 people. At the first game of the season, 32,640 fans showed up. The second game drew only 17,456. The third game brought in only 16,540 fans, although 9,500 tickets were sold before game time. What was the total attendance at these first 3 games?

18. Kris's pickup can carry 3,500 pounds of gravel in one load. On Tuesday, she carried 3 loads: one of 2,335 pounds, one of 3,476 pounds, and one of 2,890 pounds. What total weight of gravel did Kris carry in her 2 heaviest loads on Tuesday?

3 Subtraction Skills

Concepts in Subtraction

To subtract is to take things away. Subtraction, like addition, is something that is already part of your daily life. Here are two examples of subtraction:

> You pay for $6.98 worth of groceries with a ten-dollar bill, and the clerk gives you $3.02 in change.

> At a "$20 off" sale, you subtract $20 from the original price of a $49.95 coat, and you pay only $29.95.

To begin this chapter, here are five other concepts that will help you improve your subtraction skills.

1. In symbols we represent subtraction by the minus sign (−). The words *seven subtract three* are written in symbols as 7 − 3. The words *minus* and *take away* are often used in place of the word *subtract*. You can write a subtraction problem in a column or in a row. The number being subtracted from is called the **minuend**. The number being subtracted is called the **subtrahend**. The answer is called the **difference**.

   Written in a column

   $$9 \leftarrow \text{minuend}$$
   $$\underline{-5} \leftarrow \text{subtrahend}$$
   $$4 \leftarrow \text{difference}$$

   Written in a row

   $$9 - 5 = 4$$
   minuend — subtrahend — difference

2. As in addition, only like things can be subtracted. You can take one amount of money away from a larger amount. You can take a length of string from a longer piece of string. But you can't take a piece of string from a sum of money!

You can subtract like things.

You cannot subtract unlike things.

3. You cannot change the order of the minuend and subtrahend in a subtraction problem.

   12 − 7 is not the same as 7 − 12.

   Taking 35 from 100 is not the same as taking 100 from 35.

4. Subtracting zero from a number does not change the value of that number.

$$
\begin{array}{cccc}
4 & 26 & \$9 & \$15.75 \\
-0 & -\ 0 & -\ 0 & -\ \ 0.00 \\
\hline
4 & 26 & \$9 & \$15.75
\end{array}
$$

5. Word phrases that indicate subtraction are most easily solved when written as columns of numbers.

   For example, look at how these word phrases are written:

Word Phrases	Mathematical Symbols
a) two hundred ten **subtract** forty-two	210 − 42
b) the **difference** between $23.45 and $12.00	$23.45 − 12.00
c) fifty pounds **minus** thirty-five pounds	50 pounds − 35 pounds
d) nineteen feet **take away** eight feet	19 feet − 8 feet

Solve each problem as indicated.

1.
$$
\begin{array}{cccccc}
9 & 7 & 13 & 25 & \$12 & \$27.35 \\
-0 & -0 & -\ 0 & -\ 0 & -\ \ 0 & -\ \ 0.00
\end{array}
$$

2. Does the difference of twelve subtract seven equal the difference of seven subtract twelve?

Express each phrase below in mathematical symbols. Attach labels where given. You do not need to solve the problems.

3. thirty subtract twelve

4. the difference of nineteen and six

5. one hundred forty-two take away seventy-seven

6. eighty-seven pounds subtract sixty pounds

7. the difference of fifty-nine and thirty-three

8. ninety-seven feet minus forty feet

Basic Subtraction Facts

On these next two pages, you'll practice subtracting small numbers. These problems are the basic subtraction facts that you'll use in all subtraction problems, even when subtracting large numbers. Practice until you can do each problem with confidence.

Minuend (top number) smaller than 10. Subtract.

1.	8 −6	5 −2	9 −7	8 −4	6 −5	4 −0	7 −3	3 −3

2.	7 −0	9 −5	8 −5	7 −5	8 −0	7 −6	9 −3	9 −6

3.	1 −0	5 −5	8 −3	4 −1	9 −8	7 −7	5 −4	2 −0

4.	8 −2	8 −7	7 −4	6 −6	9 −0	6 −4	7 −2	4 −2

Larger Minuends. Subtract.

5.	12 − 7	14 − 7	10 − 5	13 − 4	11 − 7	10 − 7	15 − 8	11 − 5

6.	11 − 9	14 − 9	13 − 7	13 − 6	18 − 9	10 − 1	14 − 5	13 − 8

7.	11 − 6	11 − 8	10 − 4	12 − 9	10 − 3	17 − 9	10 − 6	14 − 8

8.	10 − 8	13 − 9	11 − 4	15 − 7	13 − 6	12 − 8	17 − 8	13 − 5

9.	18 − 9	15 − 6	16 − 7	14 − 6	16 − 8	15 − 9	16 − 9	12 − 6

10. In each matching exercise below, match the problems in both columns that have the same answer. Write the matching letter on the line before the number.

A. ____ 1. 9 − 4   a. 9 − 2   B. ____ 1. 16 − 8   a. 16 − 9

____ 2. 8 − 2   b. 8 − 4   ____ 2. 11 − 9   b. 12 − 8

____ 3. 6 − 4   c. 7 − 7   ____ 3. 10 − 9   c. 14 − 9

____ 4. 8 − 1   d. 4 − 1   ____ 4. 11 − 7   d. 7 − 5

____ 5. 5 − 5   e. 5 − 4   ____ 5. 14 − 8   e. 11 − 8

____ 6. 7 − 3   f. 7 − 2   ____ 6. 14 − 7   f. 1 − 0

____ 7. 8 − 5   g. 9 − 7   ____ 7. 10 − 5   g. 12 − 6

____ 8. 6 − 5   h. 9 − 3   ____ 8. 12 − 9   h. 17 − 9

Write each word phrase in numbers and labels and then solve. The first problem is done for you.

11. six buttons subtract three buttons

twelve subtract four

the difference between ten cents and five cents

6 buttons
−3 buttons
3 buttons

12. seven minus three

fourteen videotapes subtract six videotapes

twelve take away seven

13. eleven dollars take away seven dollars

fifteen cars subtract nine cars

the difference between nine and seven

14. eighteen tires minus nine tires

eleven subtract four

thirteen dollars minus eight dollars

Subtracting Larger Numbers

After learning the basic subtraction facts, you are ready to subtract larger numbers. Larger numbers are subtracted one column at a time. First you subtract the ones column at the right. Then you move to the left and subtract each column. Continue until you have subtracted each column.

To *check* a subtraction problem, add the difference (answer) to the subtrahend (the bottom number). This sum should equal the minuend (top number).

Example: Subtract: $\begin{array}{r} 745 \\ -321 \\ \hline 424 \end{array}$ Check: $\begin{array}{r} 321 \text{ (subtrahend)} \\ +424 \text{ (difference)} \\ \hline 745 \checkmark \end{array}$

Step 1. Subtract the ones column: $5 - 1 = 4$

Step 2. Subtract the tens column: $4 - 2 = 2$

Step 3. Subtract the hundreds column: $7 - 3 = 4$

Subtract. Check each answer on scratch paper.

1. $\begin{array}{r} 26 \\ -14 \end{array}$ $\begin{array}{r} 53 \\ -42 \end{array}$ $\begin{array}{r} 44 \\ -10 \end{array}$ $\begin{array}{r} 39 \\ -18 \end{array}$ $\begin{array}{r} 28 \\ -13 \end{array}$ $\begin{array}{r} 51 \\ -40 \end{array}$ $\begin{array}{r} 84 \\ -51 \end{array}$

2. $\begin{array}{r} 77 \\ -45 \end{array}$ $\begin{array}{r} 28 \\ -16 \end{array}$ $\begin{array}{r} 49 \\ -27 \end{array}$ $\begin{array}{r} 36 \\ -23 \end{array}$ $\begin{array}{r} 54 \\ -32 \end{array}$ $\begin{array}{r} 98 \\ -57 \end{array}$ $\begin{array}{r} 37 \\ -20 \end{array}$

3. $\begin{array}{r} 284 \\ -161 \end{array}$ $\begin{array}{r} 746 \\ -432 \end{array}$ $\begin{array}{r} 482 \\ -270 \end{array}$ $\begin{array}{r} 763 \\ -542 \end{array}$ $\begin{array}{r} 928 \\ -317 \end{array}$ $\begin{array}{r} 274 \\ -163 \end{array}$ $\begin{array}{r} 149 \\ -106 \end{array}$

As you subtract in the problems below, treat each empty space in a column as a zero.

4. $\begin{array}{r} 18 \\ -5 \end{array}$ $\begin{array}{r} 29 \\ -7 \end{array}$ $\begin{array}{r} 13 \\ -2 \end{array}$ $\begin{array}{r} 38 \\ -6 \end{array}$ $\begin{array}{r} 27 \\ -7 \end{array}$ $\begin{array}{r} 46 \\ -5 \end{array}$ $\begin{array}{r} 17 \\ -4 \end{array}$

5. $\begin{array}{r} 175 \\ -34 \end{array}$ $\begin{array}{r} 263 \\ -51 \end{array}$ $\begin{array}{r} 439 \\ -14 \end{array}$ $\begin{array}{r} 154 \\ -20 \end{array}$ $\begin{array}{r} 363 \\ -52 \end{array}$ $\begin{array}{r} 596 \\ -74 \end{array}$ $\begin{array}{r} 639 \\ -27 \end{array}$

6. $\begin{array}{r} 2,476 \\ -53 \end{array}$ $\begin{array}{r} 7,846 \\ -31 \end{array}$ $\begin{array}{r} 3,855 \\ -42 \end{array}$ $\begin{array}{r} 3,495 \\ -263 \end{array}$ $\begin{array}{r} 2,756 \\ -505 \end{array}$ $\begin{array}{r} 3,745 \\ -321 \end{array}$ $\begin{array}{r} 1,485 \\ -270 \end{array}$

Writing Zeros in the Answer

Sometimes when you subtract a column, you get a zero. Except for a zero in the far left column, be sure to write any zero in an answer. Remember, zero holds a place and is part of an answer.

Example 1:

$$
\begin{array}{r}
567 \\
-465 \\
\hline
102
\end{array}
$$

As part of the number, zero gives us an answer of 102, *not* 12.

Example 2:

$$
\begin{array}{r}
234 \\
-213 \\
\hline
21
\end{array}
$$

↳ In this far left column, zero is not written in the answer.

Subtract.

1.

37	46	37	12	59	35	48
−17	−26	−31	− 2	−53	− 5	−18

2.

834	927	184	750	254	583	689
−532	−321	− 82	−340	− 53	−542	− 79

3.

4,628	1,735	2,373	7,386	1,374	5,275	3,747
− 615	− 535	− 270	−2,324	−1,251	−2,240	−1,347

4.

6,384	4,593	3,585	12,474	18,595	23,274	16,648
−5,281	−2,490	−3,005	− 1,471	− 7,543	−13,150	−15,433

Find the answer to each word question below.

5. What is one hundred seventy minus sixty?

6. How many miles are left when you subtract 296 miles from 598 miles?

7. How much is left when you take 65¢ from 95¢?

8. What is the difference between two hundred fifty pounds of gravel and two hundred twenty pounds of gravel?

Subtracting Dollars and Cents

To subtract dollars and cents, line up the decimal points and then subtract the columns as you did when subtracting whole numbers. Line up the decimal points by placing one directly below the other.

Example: Subtract $2.43 from $5.89.

Step 1.   Write the problem in columns and line up the decimal points. Place $2.43 directly below $5.89.

Step 2.   From right to left, subtract each column.
Pennies column: 9 − 3 = 6
Dimes column: 8 − 4 = 4
Dollars column: 5 − 2 = 3
Bring down the decimal point.

Answer: $3.46

```
         ┌dollars column
         │ ┌dimes column
         │ │ ┌pennies column
         ↓ ↓ ↓
    $5.89
  − 2.43
    $3.46
         ↑ Decimal points
         └ are lined up.
```

Subtract. Remember to write the decimal point and dollar sign in each answer.

1.
$5.89	$12.64	$1.98	$1.75	$5.99	$0.86
− 3.25	− 9.30	− 0.75	− 1.50	− 3.45	− 0.60

2.
$27.85	$36.95	$19.75	$15.66	$95.87	$12.79
− 13.50	− 5.75	− 7.50	− 12.45	− 82.75	− 8.58

Solve each problem as indicated.

3.   Subtract $2.57 from $14.88.

4.   Subtract 75¢ from $3.99.

5.   Subtract $5.00 from $9.99.

6.   Subtract 47¢ from $1.98.

7.   Subtract $124.50 from $275.85.

8.   Subtract 37¢ from $0.98.

Subtraction Word Problems

On this page are problems that give you practice in applying subtraction skills. In each problem, watch for words that let you know you should subtract to find the answer. Examples are words such as *difference*, *left*, and *reduction*.

Before you solve each problem, read the question carefully and make sure you understand what it asks you to find.

1. Evelyn was born June 3, 1941. How old was she on June 3, 1986?

2. The price of a wool skirt was reduced from $39.99 to $25.50 during the after-Christmas sale. How much is saved with this sale price?

3. George sawed 5 inches on a board that measured 19 inches. How much is left of the longer board?

4. In Prineville, the highest recorded temperature on January 1 is 67°; the lowest recorded temperature is 13°. Find the difference between these two temperatures.

5. Before he started his diet, Stephen weighed 197 pounds. He now weighs 161 pounds. How much weight has Stephen lost?

6. Lucy takes home $895 each month in her job as a bank teller. How much money does she have left over for other expenses if she pays $375 each month for rent?

7. The distance between Kansas City and St. Louis is 228 miles. How many more miles did Mrs. Moore have to drive after driving 115 miles of this distance?

8. It is 3:46 P.M. now. What time was it thirty minutes ago?

Recognizing Key Words

A word problem may contain clues that tell you whether to add or subtract. Often, these clues are words found in the question. Certain words suggest addition, while other words suggest subtraction. These clue words are called **key words**.

Look at the use of key words in the following examples:

Example 1: Alice earned $47 on Thursday. On Friday, she earned another $36. How much *altogether* did Alice earn on Thursday *and* Friday?

Add: $47
+ 36
Answer: $83

addition key words: altogether, and

Example 2: Alice earned $47 on Thursday. On Friday, she earned another $36. How much *more* did Alice earn on Thursday *than* on Friday?

Subtract: $47
− 36
Answer: $11

subtraction key words: more . . . than

The two examples differ only in the way the questions are worded:

- In example 1, the words *altogether* and *and* suggest putting things together or adding. Example 1 is an addition problem.

- In example 2, the words *more . . . than* suggest finding a difference. To find a difference is to subtract to see how much larger one thing is than another. Example 2 is a subtraction problem.

Notice that in both examples the questions start with the words *how much*. Words like *how much, how many,* and *what* are common words used to start all types of word problems. They are not key words.

Listed below are the most common addition and subtraction key words.

Addition Key Words	*Subtraction Key Words*
add	change (money received)
and	decrease
altogether	difference
both	left
combined	less than
in all	lose (or lost)
increased	more . . . than
more	reduce (or reduction)
plus	remain (or remaining)
sum	nearer ⎫ (and other -er
total	farther ⎭ comparison words)

Underline a key word in each problem below. Then, on the line below each problem, write which operation (addition or subtraction) the key words suggest you use to solve the problem. DO NOT SOLVE ANY PROBLEM. The first one is done as an example.

1. How much change did Virginia receive when she paid a $26.45 grocery bill with a check for $30.45?

   subtraction

2. After he received his income tax refund of $235.89, Alan bought a lawn mower for $129 and several garden tools for $30. How much did Alan spend in all?

   _____

3. While driving to San Francisco, Gerry passed a sign that read "San Francisco 187 miles." After she had driven another 60 miles, how much farther did Gerry need to drive?

   _____

4. What is the difference in price between a $57,945 Mercedes Benz and an $11,230 Honda?

   _____

5. Peter weighs 184 pounds, much more than his wife, Trixie, who weighs 130 pounds. If they stood on the scale together, what would be their combined weight?

   _____

6. On a local issue during the fall election, 15,657 people voted yes. Only 6,542 people voted no. How many more yes votes were cast than no votes?

   _____

7. To save energy, Erin lowered her thermostat setting from 73 degrees to 68 degrees. By how many degrees did Erin reduce the temperature setting?

   _____

8. The Wilsons' rent of $347.50 per month is being increased by $22.50 next month. What will be their new rent total?

   _____

9. Frank's shift ended at 8:15. He stayed at work 45 more minutes to finish cleaning the floors. What time did Frank leave work?

   _____

10. Juan and Tosca both celebrate their birthdays on May 29. If Juan was born in 1946, and Tosca was born in 1959, how much older is Juan than Tosca?

    _____

Solving Word Problems

Solve each problem by adding or subtracting. In each problem, pay close attention to what you are asked to find. Be sure to look for any key words that can help you decide whether to add or to subtract.

1. The price of a $395 washing machine was reduced to $320 during the Memorial Day sale. How much can Jenny save by buying the machine at this reduced price?

2. Sharon paid for a 64¢ notepad with a fifty-cent piece and a quarter. Find how much change she should receive.

3. Travis has $575.43 in his savings account at First State Bank. If he adds $325 to this account, what will be his new savings balance?

4. At a garage sale, Mrs. Jacobs bought a chair for $17, a table for $47.25, and a lamp for $1.75. What total amount did she pay for these items?

5. When Jill's car engine developed a knocking sound, there was a total of 57,863 miles on the mileage indicator. How many more miles is this than are covered on her 50,000-mile warranty?

6. Robert's bills this month include rent of $435, a car payment of $125, gas and electricity of $147.34, and medical expenses of $67.87. What amount do these bills add up to?

7. During a "price war," Regional Air Lines offered a one-way nonstop flight from New York to Denver for $127.85. Southern Air Lines offered the same flight for $106, but you had to change planes in Chicago. Find the difference in price between these 2 flights.

8. At 8:00 A.M. the temperature was already 71°. By 3:20, the hottest part of the day, the temperature had risen 22°. Find the temperature at 3:20.

In each addition or subtraction problem below, underline the necessary information before adding or subtracting. Cross out the unnecessary information. Solve each problem and check your arithmetic.

9. While shopping, Pauline noticed that the price of a 12-ounce steak was $3.29 and the price of a smaller steak was $2.17. How much more expensive is the larger steak?

10. William wants to work a 40-hour week. On Tuesday, his boss increased his 25-hour per week work schedule to include an additional 6 hours on Saturday. How many hours in all does William now work each week?

11. During the "Spring Clearance Sale," Nathan was interested in buying a good used truck. When they reduced the price of a Ford pickup from $3795 to $2945, he bought it. If he made a down payment of $825, how much remained for him to pay off?

12. During the first month of her diet, René lost 8 pounds. Her weight dropped from 160 pounds to 152 pounds. During the second month, she lost an additional 7 pounds. So far in her third month, she has lost 5 more pounds. How many pounds in all has Rene lost on this diet?

13. A 3,955-pound full-size car gets 17 miles to the gallon. A 2,630-pound compact car gets more miles to the gallon. How many fewer pounds does the better-mileage car weigh?

14. On his dentist bill of $135.80, Dennis made payments of $25, $45.50, and $27.75. During this time, a late charge of $3.75 was added to his bill. So far, what sum has Dennis paid on his total bill?

15. By air, Chicago is 802 miles from New York City and 996 miles from Denver. Also by air, Denver is 1771 miles from New York City. How much closer is New York City to Chicago than it is to Denver? (Hint: Fill in mileage on the diagram to the right.)

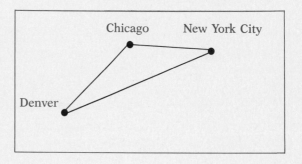

Subtracting by Borrowing

When a digit in the bottom number is larger than the digit above it, you have to **borrow** in order to subtract. The following example shows how to borrow from the tens column in order to subtract the ones column.

Example: Subtract: 53   Solution: $\overset{4\ 13}{\cancel5\ \cancel3}$ Check: $\overset{1}{2}7$
              -27 $-2\ 7$ $+26$
                                 $\overline{2\ 6}$ $\overline{53}$ ✔

Step 1. Since you can't subtract 7 from 3, borrow from the tens column. Take 1 ten from the tens column: cross out the 5 and write a 4 above it.

$\overset{4}{\cancel5}3$ 5 is replaced by 4. (1 ten is borrowed.)

Step 2. Add the borrowed 10 to the 3. Think of this as putting 13 ones in the ones column. (10 + 3 = 13 ones.)

$\overset{4\ \ 13}{\cancel5\ \cancel3}$ The borrowed 10 is added to the 3 to give 13 ones.

Step 3. Subtract the ones column: 13 − 7 = 6
Subtract the tens column: 4 − 2 = 2

$\overset{4\ \ 13}{\cancel5\ \cancel3}$ Each column can now be subtracted.
$-2\ 7$
Answer: $\overline{\textbf{2 6}}$

Subtract. Check each answer on scratch paper. Complete the partially worked Skill Builders that begin each group of rows.

Borrowing from the Tens Column

	Skill Builders						
1.	$\overset{1\ 13}{\cancel2\ \cancel3}$	$\overset{2\ 15}{\cancel3\ \cancel5}$	$\overset{7\ 14}{\cancel8\ \cancel4}$	$\overset{2\ 14}{\cancel3\ \cancel4}$	$\overset{2\ 18}{1\cancel3\ \cancel8}$	$\overset{4\ 13}{1\cancel5\ \cancel3}$	$\overset{5\ 15}{3\cancel6\ \cancel5}$
	$-\ \ 8$	$-\ \ 7$	$-5\,6$	$-1\,7$	$-\ \ 2\,9$	$-\ \ 3\,7$	$-12\,9$
	$\overline{\ \ 5}$		$\overline{\ \ 8}$		$\overline{\ \ 9}$		$\overline{\ \ 6}$

2.	47	20	62	31	26	43	83
	$-\ 9$	$-\ 6$	$-\ 5$	$-\ 4$	-18	-26	-49

3.	42	197	248	786	472	671	866
	-25	$-\ 58$	$-\ 29$	$-\ 69$	-107	-324	-358

Borrowing from the Hundreds Column

To subtract the tens column, you may need to borrow 10 tens (1 hundred) from the hundreds column.

Example:

$$\begin{array}{r} \overset{3}{\cancel{4}}\overset{11}{\cancel{1}}9 \\ -73 \\ \hline 346 \end{array}$$

Skill Builders

4.

$\overset{2}{\cancel{3}}\overset{12}{\cancel{2}}6$	$\overset{7}{\cancel{8}}\overset{12}{\cancel{2}}8$	$\overset{4}{\cancel{5}}\overset{13}{\cancel{3}}9$	$\overset{5}{\cancel{6}}\overset{13}{\cancel{3}}7$	$\overset{4}{\cancel{5}}\overset{14}{\cancel{4}}9$	$\overset{1}{\cancel{2}}\overset{15}{\cancel{3}}6$	$\overset{4}{\cancel{5}}\overset{17}{\cancel{7}}8$
-83	-73	-51	-264	-250	-184	-392
43	5		73	9	2	

5.

148	347	426	310	953	846	364
-92	-73	-84	-60	-82	-75	-80

6.

253	934	373	744	315	407	566
-172	-651	-290	-463	-261	-284	-391

Borrowing from the Thousands Column

To subtract the hundreds column, it may be necessary to borrow 10 hundreds (1 thousand) from the thousands column.

Example:

$$\begin{array}{r} \overset{1}{\cancel{2}},\overset{13}{\cancel{3}}86 \\ -874 \\ \hline 1,512 \end{array}$$

Skill Builders

7.

$\overset{3}{\cancel{4}},\overset{16}{\cancel{6}}75$	$\overset{1}{\cancel{2}},\overset{14}{\cancel{4}}75$	$\overset{0}{\cancel{1}},\overset{13}{\cancel{3}}85$	$\overset{4}{\cancel{5}},\overset{16}{\cancel{6}}82$	$\overset{7}{\cancel{8}},\overset{17}{\cancel{7}}58$	$\overset{3}{\cancel{4}},\overset{16}{\cancel{6}}83$
-943	-862	-950	$-3,871$	$-4,824$	$-3,950$
732	13	5	811	34	3

8.

5,486	4,583	3,572	7,484	1,373	2,437
-675	-720	-722	-943	-650	-525

9.

6,347	5,375	2,376	4,286	15,297	12,389
$-4,725$	$-2,840$	$-1,752$	$-2,524$	$-4,875$	$-10,850$

Borrowing from Two or More Columns

Sometimes you have to borrow more than once in a single problem. As the example shows, you do this by borrowing from column to column. Be careful to keep track of what number belongs at the top of each column.

Example: Subtract: 935   Subtract: $\overset{8\ \ 12\ 15}{\not9\ \not3\ \not5}$
                        -257 $-2\ 5\ 7$
                        ___   _____
                                                   6 7 8

Ones column: You can't subtract 7 from 5. So, borrow 1 ten from the tens column. Change the 3 to 2 and write 15 in place of 5. Subtract 7 from 15.

$$9\ \overset{2\ \ 15}{\not3\ \not5}$$
$$-2\ |5|7$$
$$8$$

Tens column: You can't subtract 5 from 2. Now borrow 1 hundred from the hundreds column. Change the 9 to 8 and write 12 in place of 2. Subtract 5 from 12.

$$\overset{8\ \ 12\ 15}{\not9\ \not3\ \not5}$$
$$-2|5|7$$
$$7|8$$

Hundreds column: Subtract 2 from 8.

$$\overset{8\ \ 12\ 15}{\not9\ \not3\ \not5}$$
$$-2|5|7$$
$$\textbf{Answer:}\quad 6|7\ 8$$

Subtract, and check each problem on scratch paper. Complete the partially worked Skill Builders that begin each group of rows.

Borrowing from the Tens and Hundreds Column

Skill Builders

1.
$\overset{3\ \ 17\ 13}{\not4\ \not8\ \not3}$ $\quad$ $\overset{5\ \ 12\ 15}{\not6\ \not3\ \not5}$ $\quad$ $\overset{5\ \ 16}{7\not6\ \not6}$ $\quad$ $\overset{1\ \ 14}{3\not2\ \not4}$ $\quad$ $\overset{8\ \ 13\ 18}{\not9\ \not4\ \not8}$ $\quad$ $\overset{4\ \ 12\ 17}{\not5\ \not3\ \not7}$ $\quad$ $\overset{1\ \ 10}{6\not2\ \not0}$

$\quad-\ \ 96\quad\quad-\ \ 67\quad\quad-\ \ 88\quad\quad-\ \ 39\quad\quad-359\quad\quad-268\quad\quad-59\ 5$

$\quad\ \ \ \overline{8\ 7}\quad\quad\ \ \ \overline{\ \ 8}\quad\quad\ \ \ \ \overline{\ \ 8}\quad\quad\quad\quad\quad\quad\ \ \overline{8\ 9}\quad\quad\quad\ \ \overline{\ \ 9}\quad\quad\quad\ \ \ \overline{\ \ 5}$

2.
654	845	637	310	327	115	845
− 96	− 79	− 38	− 28	− 39	− 47	− 86

3.
882	744	470	587	320	762	555
−595	−358	−295	−198	−267	−175	−367

Borrowing from One or More Columns: Mixed Practice

5.
26	31	63	46	237	416	142
− 9	− 8	−27	−38	− 65	− 87	− 74

6.
364	425	855	7,354	2,750	4,431
−281	−178	−367	− 632	− 820	− 650

7.
3,475	5,036	3,486	1,084	4,572	4,573
− 796	− 847	− 908	− 436	− 785	− 685

8.
4,826	2,284	3,084	6,475	12,346	34,076
−2,940	−1,379	−1,592	−3,887	− 7,209	−19,558

Subtracting from Zeros

Subtraction problems often have zeros in the top number. Since you can't borrow from a zero, you must borrow from the next column. The following example shows how to borrow from the hundreds column when there is a zero in the tens column.

Example: Subtract: 503   Solution:
$$
\begin{array}{r}
\overset{4}{\cancel{5}}\,\overset{\overset{9}{\cancel{10}}}{\cancel{0}}\,\overset{13}{\cancel{3}} \\
-2\,3\,8 \\
\hline
2\,6\,5
\end{array}
$$

$$
\begin{array}{r}
-238
\end{array}
$$

Step 1.   Since you can't subtract 8 from 3, you must borrow. But you can't borrow from the zero. You borrow from the 5. Cross out the 5 and write a 4 above it. Add the borrowed 1 hundred to the tens column. This puts 10 tens in the tens column.

$$
\begin{array}{r}
\overset{4}{\cancel{5}}\,\overset{10}{\cancel{0}}\,3 \\
-2\,3\,8
\end{array}
$$

Step 2.   Now borrow from the 10 tens. Cross out the 10 and write a 9 above it. Add the borrowed 1 ten to the ones column. This puts 13 ones in the ones column.

$$
\begin{array}{r}
\overset{4}{\cancel{5}}\,\overset{\overset{9}{\cancel{10}}}{\cancel{0}}\,\overset{13}{\cancel{3}} \\
-2\,3\,8
\end{array}
$$

Step 3.   Subtract each column:
Ones column: $13 - 8 = 5$
Tens column: $9 - 3 = 6$ **Answer:**
Hundreds column: $4 - 2 = 2$

$$
\begin{array}{r}
\overset{4}{\cancel{5}}\,\overset{\overset{9}{\cancel{10}}}{\cancel{0}}\,\overset{13}{\cancel{3}} \\
-2\,3\,8 \\
\hline
\mathbf{2\,6\,5}
\end{array}
$$

Subtract, and check each problem on scratch paper. Complete the partially worked Skill Builders in the first row.

Skill Builders						
1.	$\begin{array}{r} \overset{3}{\cancel{4}}\,\overset{\overset{9}{\cancel{10}}}{\cancel{0}}\,\overset{13}{\cancel{3}} \\ -\ \ 47 \\ \hline 6 \end{array}$	$\begin{array}{r} \overset{4}{\cancel{5}}\,\overset{\overset{9}{\cancel{10}}}{\cancel{0}}\,\overset{11}{\cancel{1}} \\ -\ \ 39 \\ \hline 2 \end{array}$	$\begin{array}{r} \overset{1}{\cancel{2}}\,\overset{\overset{9}{\cancel{10}}}{\cancel{0}}\,\overset{16}{\cancel{6}} \\ -167 \\ \hline 9 \end{array}$	$\begin{array}{r} \overset{7}{\cancel{8}}\,\overset{\overset{9}{\cancel{10}}}{\cancel{0}}\,\overset{13}{\cancel{3}} \\ -549 \\ \hline \end{array}$	$\begin{array}{r} 3,\overset{3}{\cancel{4}}\,\overset{\overset{9}{\cancel{10}}}{\cancel{0}}\,\overset{15}{\cancel{5}} \\ -\ \ 378 \\ \hline 7 \end{array}$	$\begin{array}{r} 5,\overset{4}{\cancel{5}}\,\overset{\overset{9}{\cancel{10}}}{\cancel{0}}\,\overset{13}{\cancel{3}} \\ -1,289 \\ \hline \end{array}$

2.	$\begin{array}{r} 506 \\ -\ \ 47 \\ \hline \end{array}$	$\begin{array}{r} 105 \\ -\ \ 68 \\ \hline \end{array}$	$\begin{array}{r} 230 \\ -\ \ 98 \\ \hline \end{array}$	$\begin{array}{r} 305 \\ -147 \\ \hline \end{array}$	$\begin{array}{r} 701 \\ -318 \\ \hline \end{array}$	$\begin{array}{r} 460 \\ -239 \\ \hline \end{array}$

3.	$\begin{array}{r} 503 \\ -427 \\ \hline \end{array}$	$\begin{array}{r} 902 \\ -335 \\ \hline \end{array}$	$\begin{array}{r} 7,450 \\ -\ \ 239 \\ \hline \end{array}$	$\begin{array}{r} 3,501 \\ -\ \ 843 \\ \hline \end{array}$	$\begin{array}{r} 5,305 \\ -1,236 \\ \hline \end{array}$	$\begin{array}{r} 2,502 \\ -1,825 \\ \hline \end{array}$

Subtracting from Separated Zeros

The following example shows how to use borrowing to subtract from numbers that have separated zeros.

Example: Subtract:
$$\begin{array}{r} 2{,}0\,6\,0 \\ -1{,}3\,7\,9 \end{array}$$

Solution:
$$\begin{array}{r} \overset{9\ 15}{\overset{1\ \,10\ \,\cancel{8}\,10}{\cancel{2}{,}\cancel{0}\,\cancel{6}\,\cancel{0}}} \\ -1{,}3\,7\,9 \\ \hline 6\,8\,1 \end{array}$$

Step 1. The first nonzero digit in the subtraction problem is 6. Cross out the 6 and write 5 above it. Change the first zero to 10.

$$\overset{\quad\ \ 5\ \,10}{2{,}0\,\cancel{6}\,\cancel{0}}$$

Step 2. The second nonzero digit is a 2. Cross out the 2 and write 1 above it. Change the second zero to a 10.

$$\overset{1\ \ 10\ \,5\ \,10}{\cancel{2}{,}\cancel{0}\,\cancel{6}\,\cancel{0}}$$

Step 3. Subtract. Since you can't take 7 from 5, borrow from the 10.

$$\begin{array}{r} \overset{9\ 15}{\overset{1\ \,10\ \,\cancel{8}\,10}{\cancel{2}{,}\cancel{0}\,\cancel{6}\,\cancel{0}}} \\ -1{,}3\,7\,9 \end{array}$$

Answer: $6\,8\,1$

Complete each subtraction problem below. In the Skill Builder row, we have worked out some of the necessary borrowing. For practice, see if you can follow how each borrowing step was completed. Use some scratch paper, if you'd like.

Skill Builders

1.
$$\begin{array}{r}\overset{9\ 17}{\overset{4\ \,10\ \,\cancel{7}\,10}{\cancel{5}{,}\cancel{0}\,\cancel{8}\,\cancel{0}}} \\ -\ \ 7\,9\,3 \\ \hline 8\,7 \end{array}$$
$$\begin{array}{r}\overset{\quad\ \ 5\ \,10}{4{,}0\,\cancel{6}\,\cancel{0}} \\ -\ \ 8\,7\,5 \\ \hline 5 \end{array}$$
$$\begin{array}{r}\overset{9\ 12}{\overset{6\ \,10\ \,\cancel{7}\,10}{\cancel{7}{,}\cancel{0}\,\cancel{3}\,\cancel{0}}} \\ -4{,}8\,4\,6 \\ \hline 8\,4 \end{array}$$
$$\begin{array}{r}\overset{\quad\ \ 7\ \,10}{2{,}0\,\cancel{8}\,\cancel{0}} \\ -1{,}3\,5\,9 \\ \hline 2\,1 \end{array}$$

2.
$$\begin{array}{r} 4{,}0\,9\,0 \\ -\ \ 6\,3\,8 \end{array}$$
$$\begin{array}{r} 7{,}0\,5\,0 \\ -\ \ 7\,4\,5 \end{array}$$
$$\begin{array}{r} 3{,}0\,1\,0 \\ -\ \ 1\,9\,3 \end{array}$$
$$\begin{array}{r} 8{,}0\,7\,0 \\ -\ \ 6\,7\,4 \end{array}$$
$$\begin{array}{r} 5{,}0\,4\,0 \\ -2{,}7\,3\,3 \end{array}$$

Subtracting from a Row of Zeros

In the following example, there is more than one zero in a row in the top number. To subtract, you borrow from the first nonzero digit to the left.

Example: Subtract: $\begin{array}{r} 5{,}000 \\ -2{,}165 \end{array}$ Solution: $\begin{array}{r} \overset{4\ \ \overset{9}{\cancel{10}}\overset{9}{\cancel{10}}10}{\cancel{5}{,}\cancel{0}\,\cancel{0}\,\cancel{0}} \\ -2{,}1\ 6\ 5 \\ \hline 2{,}8\ 3\ 5 \end{array}$

Step 1. The first nonzero digit is 5. Cross out the 5 and write 4 above it. Change the first zero to 10.

$\overset{4\ \ 10}{\cancel{5}{,}\cancel{0}}\,0\,0$

Step 2. Cross out the 10 and write 9 above it. Change the next 0 to 10.

$\overset{4\ \ \overset{9}{\cancel{10}}\,10}{\cancel{5}{,}\cancel{0}\,\cancel{0}}\,0$

Step 3. Cross out this 10 and write 9 above it. Change the last 0 to 10. Subtract each column.

$\begin{array}{r} \overset{4\ \ \overset{9}{\cancel{10}}\overset{9}{\cancel{10}}10}{\cancel{5}{,}\cancel{0}\,\cancel{0}\,\cancel{0}} \\ -2{,}1\ 6\ 5 \end{array}$

Answer: 2,8 3 5

Note: In this type of problem, all zeros in a row (except the 0 in the ones column) end up as nines. Knowing this can save you a lot of time in many problems.

Skill Builders

1.

$\overset{2\ \overset{9}{\cancel{10}}10}{\cancel{3}\,\cancel{0}\,\cancel{0}}$	$\overset{5\ \overset{9}{\cancel{10}}10}{\cancel{6}\,\cancel{0}\,\cancel{0}}$	$\overset{6\ \overset{9}{\cancel{10}}10}{\cancel{7}\,\cancel{0}\,\cancel{0}}$	$\overset{4\ \overset{9}{\cancel{10}}\overset{9}{\cancel{10}}10}{\cancel{5}{,}\cancel{0}\,\cancel{0}\,\cancel{0}}$	$\overset{2\ \overset{9}{\cancel{10}}\overset{9}{\cancel{10}}10}{\cancel{3}{,}\cancel{0}\,\cancel{0}\,\cancel{0}}$	$\overset{3\ \overset{9}{\cancel{10}}\overset{9}{\cancel{10}}10}{2\cancel{4}{,}\cancel{0}\,\cancel{0}\,\cancel{0}}$
$-\ \ 89$	-353	-286	$-\ \ 487$	$-1{,}764$	$-11{,}855$
$1\ 1$	7		$1\ 3$	6	

2.

200	500	800	500	800	700
$-\ 45$	$-\ 73$	$-\ 26$	-376	-392	-453

3.

3,000	2,000	4,000	6,000	32,000	15,000
$-\ \ 840$	$-\ \ 950$	$-2{,}405$	$-3{,}775$	$-16{,}740$	$-\ 8{,}969$

Subtracting Numbers in a Row

When numbers are written in a row, rewrite them in a column before subtracting. The smaller number is placed directly below the larger number. Be sure to place the ones digits in a ones column, tens digits in a tens column, and so on. Remember to borrow when you cannot subtract otherwise.

Example: Subtract: $1,507 - 435$

Place 435 directly below 1,507 by first lining up the ones digits. Place the 5 below the 7. The tens and hundreds digits will then line up correctly.

Solution:
$$\begin{array}{r} 1,507 \\ -435 \\ \hline \mathbf{1,072} \end{array}$$

Rewrite the following problems in a column and subtract.

1. $98 - 37$ $483 - 231$ $2,678 - 1,342$

2. $187 - 69$ $226 - 192$ $3,265 - 1,741$

3. $364 - 89$ $3,486 - 798$ $506 - 357$

4. $1,060 - 427$ $200 - 89$ $400 - 173$

5. $800 - 247$ $1,000 - 429$ $12,000 - 7,736$

Borrowing with Dollars and Cents

Borrowing with dollars and cents is done the same way as borrowing with whole numbers. You borrow from one column to the next as if the decimal point weren't there. When you are done subtracting, be sure to bring down the decimal point.

Example: Subtract: $14.00   Solution: $\overset{3\ \overset{9}{\cancel{10}}\ 10}{\$14.\cancel{0}\ \cancel{0}}$ Check: $\overset{1\ \ 1}{3.27}$
                      $-\quad 3.27$ $-\quad 3.2\ 7$ $+\$10.73$
                                                     $\$10.7\ 3$ $\$14.00$ ✓

Subtract, and check each answer on scratch paper.

1. $3.45 $6.06 $12.64 $40.82 $153.47 $235.50
    − 1.28   − 2.19   − 9.39   − 13.95   − 74.80   − 125.95

2. $5.00 $8.00 $15.00 $52.00 $300.00 $500.00
    − 2.37   − 4.58   − 7.37   − 23.83   − 142.65   − 389.99

3. $5.88 − 89¢ $1.50 − 79¢ $2.00 − 28¢ $5.00 − 63¢

Solve the two subtraction problems indicated below.

4.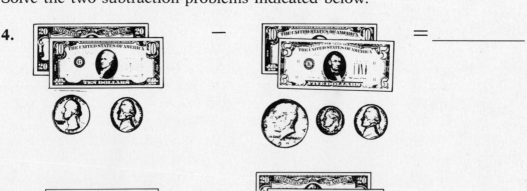

5.

Deciding When to Add and When to Subtract

How do you know when to add or when to subtract? One way is to look for key words. Unfortunately, many word problems do not contain key words. Or, in some problems, key words may not be helpful.

A better approach is to make sure you understand the problem. Then, if you're still not sure what to do, don't be afraid to solve the problem first by adding, then by subtracting. In most cases, only the one correct answer will make sense.

Example: Ted paid a total of $12.79 for a case of car oil. He gave the clerk a twenty-dollar bill. How much money did the clerk then give to Ted?

Notice that the question does not contain a key word. The only key word is the word *total* in the first sentence. Since *total* is a common addition key word, you may think you should solve this problem by adding.

Incorrectly Solved by adding
$20.00
+ 12.79
$32.79

But, what really happens when you shop? Try to picture Ted and the clerk in your mind. The clerk gives change to Ted. To get change, you subtract the cost from the amount paid. Thus, your mental picture suggests solving this problem by subtracting.

Correctly Solved by subtracting
$20.00
− 12.79
$ 7.21

A look at both answers shows which is correct. Your own experience tells you that only $7.21 makes sense. The answer $32.79 is more than the $20.00 that Ted gave to the clerk!

The example shows how you might get fooled by key words. As used here, *total* refers only to Ted's bill of $12.79. *Total* is not a clue to solving this problem. Here the most important clue is your own experience. You can best use this experience by remembering the following:

● **The final step in solving a word problem is to reread the question and make sure your answer makes sense.**

To the right of each problem, circle the answer that makes the most sense. One answer is obtained by addition; the other by subtraction. Only one answer is correct. YOU DO NOT NEED TO SOLVE THESE PROBLEMS.

1. Ann purchased a hammer for $7.49. Paying with a ten-dollar bill, how much change did she receive?

   a) addition: $17.49
   b) subtraction: $2.51

2. If Sam received $4.81 in change after paying with $20, what was the cost of his purchase?

    a) addition: $24.81
    b) subtraction: $15.19

3. On Monday the price of a new Toyota was $7,849. By Friday the price had been changed to $5,995 due to a factory rebate. What value is this rebate to a buyer?

    a) addition: $13,844
    b) subtraction: $1,854

4. The time in Seattle is 3 hours earlier than the time in New York. What time is it in New York when it is 10:00 A.M. in Seattle?

    a) addition: 1:00 P.M.
    b) subtraction: 7:00 A.M.

5. While shopping, April bought 2 gallons of paint at a total cost of $23.98. If she paid with 2 twenty-dollar bills, what amount did the clerk give her?

    a) addition: $63.98
    b) subtraction: $16.02

6. While working on construction, Jeff ate about 3,800 calories a day. When he was laid off, he ate 2,350 calories a day. How many additional calories did Jeff eat each day as a construction worker?

    a) addition: 6,150 calories
    b) subtraction: 1,450 calories

7. Jack has two packages to mail. If the smaller one costs $2.57 to send, and the larger one costs $3.74, what postage will he be charged?

    a) addition: $6.31
    b) subtraction: $1.17

8. At 5:00 A.M. the temperature was a cool 26 degrees. By 11:00 A.M. the sun had warmed everything up by 18 degrees. What was the temperature at 11:00 A.M.?

    a) addition: 44 degrees
    b) subtraction: 8 degrees

9. After changing jobs, Shelley's salary was $11,490. At her previous job, she made $13,500. What total yearly amount did Shelley give up to take this new job?

    a) addition: $24,990
    b) subtraction: $2,010

Solving Word Problems

On the next two pages are addition and subtraction problems for you to solve. Some of these problems contain helpful key words and some don't. Many of the problems contain more information than is needed for answering the question.

Solve each problem carefully and check your arithmetic. Also, check to see that your answer makes sense. If it doesn't, you may have added when you should have subtracted (or vice versa)! Redo the problem until your answer seems right.

1. Jessica paid for a 79¢ milk shake with a one-dollar bill. How much change did she get?

2. Jon's car weighs 3,145 pounds, and his boat and trailer together weigh 876 pounds. How much more does his car weigh than his boat and trailer?

3. After his wages were reduced by $125.75 per month in take-home pay, Robby's monthly check was $837.87. What was Robby's monthly check before the wage reduction?

4. The time in San Francisco is 2 hours earlier than the time in Chicago. When it is 8:00 A.M. in San Francisco, it is 10:00 A.M. in Chicago. What time is it in Chicago when it is 3:00 P.M. in San Francisco?

   8:00 AM ⟶ 10:00 AM
   3:00 PM ⟶ ?
   San Francisco   Chicago

5. During June, Ella's landlord announced that her rent would be increased in August from $350 to $429. During July, Ella moved to a new apartment where the rent was $415. How much additional rent is Ella paying now compared to what she used to pay?

6. To save heat energy in the home, it is recommended that thermostats be set no higher than 68 degrees Fahrenheit. How many degrees lower is this than the Jones family thermostat setting of 73 degrees?

7. If Sarah was born on August 6, 1915, in what year will she celebrate her 80th birthday?

8. A mile is the same as 1,760 yards. How many yards farther is a mile than a kilometer, which is about 1,100 yards?

9. Craig drove his car an average of 1,154 miles each month last year. During the first three months this year, he has driven 954 miles, 876 miles, and 1,105 miles. What number of miles has he put on his car so far this year?

10. To lose 1 pound each week, Wilma must reduce her food intake by 500 calories each day. If her normal daily diet consists of 2,350 calories, what new daily calorie level must Wilma maintain in order to lose 1 pound per week?

11. Wilma's diet chart showed the following information: Hamburger, 4 ounces = 324 calories; Fish, 4 ounces = 120 calories; Chicken, 4 ounces = 280 calories. How many more calories is 4 ounces of chicken than 4 ounces of fish?

12. When Royce and Bev went out to eat, they shared a pizza for $7.95, two salads for $2.98 each, and two soft drinks for $1.20 each. If the tax came to 48¢, what was the total price of their meal?

13. A case of oil that normally sells for $12.99 is on sale for 84¢ a quart or $9.99 a case. In addition, the oil company is offering a $2.00 rebate on each case. If Sharon buys the oil at the sale price and sends for the rebate, how much will a case of oil end up costing her?

Word problems are often written as a short story followed by more than one question. Each question requires its own necessary information. Choose this information carefully as you answer each question below.

Questions 14 through 16 refer to the story below.

Betty works as a secretary for Tillman Construction Company. Her monthly salary is $960. Her actual take-home pay after taxes is $720.

Betty's 3 main expenses each month are $275 for rent, $137 for a car payment, and $156 for food. Because she works hard and still has a tough time meeting expenses, Betty recently asked for a raise.

14. How much money is subtracted from Betty's check each month to pay taxes?

Information needed:

Solution:

15. What is the sum of Betty's 3 main monthly expenses?

Information needed:

Solution:

16. How much more does Betty pay for rent each month than she pays for food?

Information needed:

Solution:

Questions 17 through 19 refer to the following story.

Frank was overweight and decided to go on a diet. Before starting the diet, Frank ate 3,225 calories of food each day. On doctor's orders, Frank reduced his food intake by 750 calories each day. In 3 months Frank's weight decreased from 214 pounds to 196 pounds.

After losing the weight, Frank felt so good that he decided to buy a pickup truck. He found a Chevrolet on sale for $2845. An optional camper was $735. Frank bought both.

17. While on his diet, how many calories of food was Frank allowed each day?

Information needed:

Solution:

18. How many pounds did Frank lose during the 3-month period?

Information needed:

Solution:

19. What total price did Frank pay for the pickup and camper together?

Information needed:

Solution:

Subtraction Skills Review

On the next two pages, you'll have a chance to review your subtraction skills. Work each problem as carefully as you can and check your answers with the answers on page 175. Correct any mistakes.

SUBTRACTING SMALL NUMBERS: Review pages 44 through 47.

1.
$$\begin{array}{r} 9 \\ -6 \\ \hline \end{array} \qquad \begin{array}{r} 8 \\ -8 \\ \hline \end{array} \qquad \begin{array}{r} 15 \\ -\ 4 \\ \hline \end{array} \qquad \begin{array}{r} 14 \\ -\ 8 \\ \hline \end{array} \qquad 3 - 0 = \qquad 12 - 5 =$$

SUBTRACTING LARGER NUMBERS, NO BORROWING: Review pages 48 through 50.

2.
$$\begin{array}{r} 37 \\ -14 \\ \hline \end{array} \qquad \begin{array}{r} 46 \\ -30 \\ \hline \end{array} \qquad \begin{array}{r} 28 \\ -\ 7 \\ \hline \end{array} \qquad \begin{array}{r} 674 \\ -\ 54 \\ \hline \end{array} \qquad \begin{array}{r} 483 \\ -182 \\ \hline \end{array} \qquad \begin{array}{r} 3,448 \\ -1,215 \\ \hline \end{array}$$

SUBTRACTING AND BORROWING: Review pages 56 through 61.

3.
$$\begin{array}{r} 35 \\ -\ 8 \\ \hline \end{array} \qquad \begin{array}{r} 47 \\ -19 \\ \hline \end{array} \qquad \begin{array}{r} 183 \\ -\ 54 \\ \hline \end{array} \qquad \begin{array}{r} 325 \\ -118 \\ \hline \end{array} \qquad \begin{array}{r} 462 \\ -278 \\ \hline \end{array} \qquad \begin{array}{r} 1,382 \\ -\ 718 \\ \hline \end{array}$$

4.
$$\begin{array}{r} 460 \\ -168 \\ \hline \end{array} \qquad \begin{array}{r} 705 \\ -\ 39 \\ \hline \end{array} \qquad \begin{array}{r} 3,047 \\ -\ 828 \\ \hline \end{array} \qquad \begin{array}{r} 3,050 \\ -\ 439 \\ \hline \end{array} \qquad \begin{array}{r} 5,030 \\ -1,547 \\ \hline \end{array} \qquad \begin{array}{r} 9,000 \\ -2,583 \\ \hline \end{array}$$

SUBTRACTING DOLLARS AND CENTS: Review pages 50 and 64.

5.
$$\begin{array}{r} \$5.40 \\ -\ 1.30 \\ \hline \end{array} \qquad \begin{array}{r} \$3.93 \\ -\ 2.80 \\ \hline \end{array} \qquad \begin{array}{r} \$9.46 \\ -\ 4.39 \\ \hline \end{array} \qquad \begin{array}{r} \$8.38 \\ -\ 2.49 \\ \hline \end{array} \qquad \begin{array}{r} \$6.80 \\ -\ 4.63 \\ \hline \end{array} \qquad \begin{array}{r} \$20.00 \\ -\ 8.49 \\ \hline \end{array}$$

SUBTRACTING NUMBERS WRITTEN IN A ROW: Review page 63.

6. 18 − 9   36 − 8   29 − 14   146 − 27

7. 183 − 97   207 − 183   400 − 137   5,020 − 2,394

RECOGNIZING SUBTRACTION WORD PHRASES: Review pages 45, 52, and 65.

Write each word phrase in symbols and then solve.

8. twenty minus thirteen

9. the difference of forty dollars and two dollars

10. twenty-two inches subtracted from fifty-one inches

Write each amount in decimal form and then solve.

11. $0.74 subtract 19¢

12. the difference of $4.28 and 99¢

13. 72¢ take away 9¢

SOLVING WORD PROBLEMS: Review pages 52 through 55 and pages 65 through 69.

14. What is the difference between a one-dollar bill and the sum of three quarters, one dime, and four pennies?

15. If Janice paid for an 84¢ hot dog with a two-dollar bill, how much change should she receive?

16. By air, Philadelphia is 738 miles from Chicago. Also by air, Philadelphia is 668 miles from Louisville. How much nearer is Philadelphia to Louisville than it is to Chicago? (Hint: Drawing a diagram might help.)

17. George bought a lawnmower that regularly sells for $289 at a reduced price of $239. He also bought several garden tools for a total price of $37.50, a savings of $14.75. What amount did George save on the cost of the lawnmower?

18. The time in Los Angeles is 3 hours earlier than the time in New York. When it is 7:00 A.M. in Los Angeles, it is 10:00 A.M. in New York City. What time is it in Los Angeles when it is 5:00 P.M. in New York?

Los Angeles
7:00 A.M.
?

New York
10:00 A.M.
5:00 P.M.

4 Multiplication Skills

Concepts in Multiplication

Multiplication is a shortcut that takes the place of repeated addition. Instead of adding a number several times, you multiply and get the answer in one step. Here is an example:

- Multiply to find the cost of 4 tires if each tire costs $62.00.

Solved by multiplication

$ 62.00
×   4
————————
$248.00

Solved by addition

$ 62.00
62.00
62.00
+   62.00
————————
$248.00

As you begin this chapter, become familiar with the following five concepts:

1. Multiplication is represented by the times sign (×). For example, six times four is written in symbols as 6 × 4.

Symbols

6 × 4

Equivalent Word Phrases

six <u>times</u> four
six <u>multiplied by</u> four
the <u>product of</u> six and four

You may see a multiplication problem written in a column (up and down) or in a row (across). The answer is called the ***product.***

Parts of a Multiplication Problem

Written in a column

6 ← multiplicand
× 4 ← multiplier
————
24 ← product

Written in a row

6 × 4 = 24

product
multiplier
multiplicand

2. Changing the order of numbers being multiplied does not change the product.

$$\begin{array}{r} 6 \\ \times\ 4 \\ \hline 24 \end{array} \quad \text{and} \quad \begin{array}{r} 4 \\ \times\ 6 \\ \hline 24 \end{array} \qquad \begin{array}{r} 12 \\ \times\ 3 \\ \hline 36 \end{array} \quad \text{and} \quad \begin{array}{r} 3 \\ \times 12 \\ \hline 36 \end{array}$$

3. The product of any number times 1 is the same number.

$$\begin{array}{r} 5 \\ \times 1 \\ \hline 5 \end{array} \qquad \begin{array}{r} 28 \\ \times\ 1 \\ \hline 28 \end{array} \qquad \begin{array}{r} \$6.07 \\ \times\ \ \ \ 1 \\ \hline \$6.07 \end{array} \qquad \begin{array}{r} 1 \\ \times 9 \\ \hline 9 \end{array} \qquad \begin{array}{r} 1 \\ \times 30 \\ \hline 30 \end{array}$$

4. The product of any number times 0 is 0.

$$\begin{array}{r} 8 \\ \times 0 \\ \hline 0 \end{array} \qquad \begin{array}{r} 36 \\ \times\ 0 \\ \hline 0 \end{array} \qquad \begin{array}{r} \$1.70 \\ \times\ \ \ \ 0 \\ \hline 0 \end{array} \qquad \begin{array}{r} 0 \\ \times 7 \\ \hline 0 \end{array} \qquad \begin{array}{r} 0 \\ \times 92 \\ \hline 0 \end{array}$$

5. You can rewrite phrases that indicate multiplication as a multiplication problem. Look at these examples:

Word Phrases	*Mathematical Symbols*
a) thirty-seven **times** nineteen	$\begin{array}{r} 37 \\ \times 19 \\ \hline \end{array}$
b) one hundred nine **multiplied by** twelve	$\begin{array}{r} 109 \\ \times\ \ 12 \\ \hline \end{array}$

Solve each problem below as indicated.

1. $\begin{array}{r} 15 \\ \times\ 1 \\ \hline \end{array}$ $\qquad \begin{array}{r} \$.09 \\ \times\ \ \ \ 1 \\ \hline \end{array}$ $\qquad \begin{array}{r} 1 \\ \times 7 \\ \hline \end{array}$ $\qquad \begin{array}{r} 1 \\ \times 8 \\ \hline \end{array}$ $\qquad \begin{array}{r} \$1.26 \\ \times\ \ \ \ 0 \\ \hline \end{array}$ $\qquad \begin{array}{r} 0 \\ \times 7 \\ \hline \end{array}$

2. Which two of the following equals the product of 6 × 3?

   9 + 9 + 9 $\qquad$ 3 + 6 $\qquad$ 3 × 6 $\qquad$ 6 + 6 + 6

Write each phrase below in mathematical symbols. Attach labels where given. You do not need to do the multiplication.

3. two hundred forty-nine pounds multiplied by eighteen

4. forty times twenty-three

5. five dollars and eighteen cents times six

6. four hundred miles multiplied by seven

Basic Multiplication Facts

On these next two pages, you'll practice multiplying single-digit numbers. These problems are the basic multiplication facts you'll use in all multiplication. These facts are shown in the **multiplication table** below.

You can use the multiplication table to find the answer to single-digit problems. The numbers that you multiply are written in the shaded row and column. It is a good idea to memorize this table if you don't already know it.

The table does not include 0. For problems involving 0, remember that the product of any number times 0 is 0.

The Multiplication Table

	1	2	3	4	5	6	7	8	9
1	1	2	3	4	5	6	7	8	9
2	2	4	6	8	10	12	14	16	18
3	3	6	9	12	15	18	21	24	27
4	4	8	12	16	20	24	28	32	36
5	5	10	15	20	25	30	35	40	45
6	6	12	18	24	30	36	42	48	54
7	7	14	21	28	35	42	49	56	63
8	8	16	24	32	40	48	56	64	72
9	9	18	27	36	45	54	63	72	81

Example: Multiply: 6
                          $\times 7$

Step 1. Find 6 in the shaded row across the top.

Step 2. Run your finger down the 6 column until you reach the row starting with a shaded 7.

The square at the intersection contains the answer.

Answer: 42

Multiplying by Two Digits

When you multiply by a two-digit number, you multiply the top number by each digit of the bottom number:

- When you multiply by the ones digit, start your answer in the ones column.

- When you multiply by the tens digit, start your answer in the tens column.

The answer of each step of the multiplication process is called a **partial product**. The answer of the whole problem is computed by adding the partial products.

Example: Multiply: 42   Solution: 42

$$\begin{array}{r} 42 \\ \times\,13 \\ \hline \end{array} \qquad \begin{array}{r} 42 \\ \times\ 13 \\ \hline 126 \\ 42 \\ \hline 546 \end{array}$$

partial products { 126 / 42 ← Treat this blank space as a zero.

Step 1.   Multiply 42 by the ones digit: $42 \times 3 = 126$
Write 126.

Step 2.   Multiply 42 by the tens digit: $42 \times 1 = 42$
Be sure to write 42, the second partial product, in the tens and hundreds columns. It does not start in the ones column.

Step 3.   Add the partial products.

Answer: **546**

Multiply. Complete the partially worked Skill Builders in row 1.

Skill Builders

Write the second partial product in the boxes below, then add to find the answer.

1.
$$\begin{array}{r} 31 \\ \times\,22 \\ \hline 62 \\ 62 \end{array} \quad \begin{array}{r} 30 \\ \times\,23 \\ \hline 90 \\ 60 \end{array} \quad \begin{array}{r} 40 \\ \times\,12 \\ \hline 80 \\ \boxed{} \end{array} \quad \begin{array}{r} 52 \\ \times\,23 \\ \hline 156 \\ \boxed{} \end{array} \quad \begin{array}{r} 420 \\ \times\ 14 \\ \hline 1680 \\ 420 \end{array} \quad \begin{array}{r} 521 \\ \times\ 24 \\ \hline 2084 \\ \boxed{} \end{array} \quad \begin{array}{r} 422 \\ \times\ 43 \\ \hline 1266 \\ \boxed{} \end{array}$$

2.
$$\begin{array}{r} 32 \\ \times\,12 \\ \hline \end{array} \quad \begin{array}{r} 24 \\ \times\,22 \\ \hline \end{array} \quad \begin{array}{r} 31 \\ \times\,11 \\ \hline \end{array} \quad \begin{array}{r} 40 \\ \times\,21 \\ \hline \end{array} \quad \begin{array}{r} 41 \\ \times\,37 \\ \hline \end{array} \quad \begin{array}{r} 421 \\ \times\ 43 \\ \hline \end{array} \quad \begin{array}{r} 512 \\ \times\ 32 \\ \hline \end{array}$$

Multiplying by Three Digits

When you multiply by a three-digit number, you get three partial products:

- Start the first partial product in the ones column.
- Start the second partial product in the tens column.
- Start the third partial product in the hundreds column.

Example: Multiply: 312
  $\times$ 231

Solution:
```
        312
      × 231
1st →   312
2nd →   936
3rd →   624
      72,072
```
partial products

Step 1. Multiply 312 by the ones digit: 312 × 1 = 312

Step 2. Multiply 312 by the tens digit: 312 × 3 = 936

Step 3. Multiply 312 by the hundreds digit: 312 × 2 = 624

Step 4. Add the partial products. Treat all blank spaces as zeros.

Answer: 72,072

Multiply. Complete the partially worked Skill Builders in row 1.

Skill Builders

Write the second and third partial products in the boxes below.

1.
```
   132      322      310      402     1,200    1,141    2,321
 ×121      ×213     ×131     ×221     ×  312   ×  211   ×  213
   132      966      310      402      2 400    1 141    6 963
   264      322      930       □      12 00    11 41      □
   132       □        □        □      360 0      □        □
```

2.
```
   324      211      321      423      124      230      300
 ×112     ×121     ×211     ×212     ×112     ×212     ×121
```

3.
```
 1,300    2,400    3,100    1,231    3,202    3,112    4,120
 ×  122   ×  221   ×  311   ×  113   ×  212   ×  312   ×  211
```

78

Zeros in Multiplication

When multiplying by zero, you can use a shortcut. Write a 0 directly beneath the 0 in the problem. Then multiply by the next digit and start the new partial product in the next space to the left. Using the shortcut, you do not need to write a partial product that contains only 0s.

Example 1:   *Shortcut*   *Long Way*

$$\begin{array}{r} 21\,2 \\ \times\ \ 3|0 \\ \hline 6,36\,|0 \end{array}$$

Write a 0 beneath the 0 above.

$$\begin{array}{r} 21\,2 \\ \times\ \ 3|0 \\ \hline 00\,|0 \\ 6\ 36 \\ \hline 6,36\ 0 \end{array}$$

This partial product contains only 0s.

The shortcut is to write the second partial product to the left of the zero. This shortcut reduces the number of partial products you write when you multiply by 0. Look at the next two examples carefully.

Example 2:   *Shortcut*   *Long Way*

$$\begin{array}{r} 2\ 4|3 \\ \times\ 1|1\ 0 \\ \hline 2\ 4|3\ 0 \\ 24\ 3 \\ \hline 26,7|3\ 0 \end{array}$$

$$\begin{array}{r} 243 \\ \times\ 110 \\ \hline 000 \\ 2\ 43 \\ 24\ 3 \\ \hline 26,730 \end{array}$$

Example 3:   *Shortcut*   *Long Way*

$$\begin{array}{r} 3\ 2\ 1 \\ \times\ 1|0|2 \\ \hline 6\ 4\ 2 \\ 32\ 1|0 \\ \hline 32,7\ 4\ 2 \end{array}$$

$$\begin{array}{r} 321 \\ \times\ 102 \\ \hline 642 \\ 0\ 00 \\ 32\ 1 \\ \hline 32,742 \end{array}$$

Multiply.

1.
$$\begin{array}{r} 23 \\ \times 20 \end{array}$$
$$\begin{array}{r} 43 \\ \times 10 \end{array}$$
$$\begin{array}{r} 21 \\ \times 20 \end{array}$$
$$\begin{array}{r} 51 \\ \times 40 \end{array}$$
$$\begin{array}{r} 132 \\ \times\ 30 \end{array}$$
$$\begin{array}{r} 712 \\ \times\ 40 \end{array}$$
$$\begin{array}{r} 413 \\ \times\ 30 \end{array}$$

2.
$$\begin{array}{r} 321 \\ \times 210 \end{array}$$
$$\begin{array}{r} 132 \\ \times 320 \end{array}$$
$$\begin{array}{r} 212 \\ \times 210 \end{array}$$
$$\begin{array}{r} 130 \\ \times 110 \end{array}$$
$$\begin{array}{r} 220 \\ \times 140 \end{array}$$
$$\begin{array}{r} 213 \\ \times 210 \end{array}$$
$$\begin{array}{r} 312 \\ \times 120 \end{array}$$

3.
$$\begin{array}{r} 243 \\ \times 200 \end{array}$$
$$\begin{array}{r} 421 \\ \times 400 \end{array}$$
$$\begin{array}{r} 612 \\ \times 300 \end{array}$$
$$\begin{array}{r} 502 \\ \times 200 \end{array}$$
$$\begin{array}{r} 212 \\ \times 103 \end{array}$$
$$\begin{array}{r} 413 \\ \times 203 \end{array}$$
$$\begin{array}{r} 421 \\ \times 301 \end{array}$$

Multiplying by 10, 100, and 1,000

The shortcut on the previous page gives us three easy rules to use when either the top number or bottom number is 10, 100, or 1,000.

★ **To multiply a number by 10, add a 0 to the right of the number.**

Examples:

(1)	35	(2)	278	(3)	10
	× 10		× 10		× 17
	350		2,780		170

★ **To multiply a number by 100, add two 0s to the right of the number.**

Examples:

(1)	463	(2)	3,647	(3)	100
	× 100		× 100		× 89
	46,300		364,700		8,900

★ **To multiply a number by 1,000, add three 0s to the right of the number.**

Examples:

(1)	457	(2)	4,826	(3)	1,000
	× 1,000		× 1,000		× 27
	457,000		4,826,000		27,000

Multiply.

1.

25	67	59	135	352	589
× 10	× 10	× 10	× 10	× 10	× 10

2.

127	254	634	352	485	386
× 100	× 100	× 100	× 100	× 100	× 100

3.

100	10	100	10	100	10
× 85	× 9	× 76	× 7	× 365	× 14

4.

1,000	1,000	1,000	1,000	578	4,586
× 36	× 65	× 375	× 634	× 1,000	× 1,000

Multiplying Dollars and Cents

When multiplying dollars and cents, multiply as if the decimal point weren't there. Then place the decimal point and dollar sign in the answer to separate dollars from cents.

Remember, cents go in the first two places to the right of the decimal point.

Example 1:

Multiply:

$$\begin{array}{r} \$4.23 \\ \times\ \ 3 \\ \hline \$12.69 \end{array}$$

Step 1. Multiply each digit:

$$\begin{array}{r} \$4.23 \\ \times\ \ 3 \\ \hline 12\ 69 \end{array}$$

Step 2. Place the decimal point and dollar sign in the answer.

$$\begin{array}{r} \$4.23 \\ \times\ \ 3 \\ \hline \$12.69 \end{array}$$

Example 2:

Multiply:

$$\begin{array}{r} \$5.31 \\ \times\ 10 \\ \hline \$53.10 \end{array}$$

Step 1. To multiply by 10, add one 0 to the number:

53 10

Step 2. Place the decimal point in the answer to separate dollars from cents. Write the dollar sign.

$53.10

Multiply. Remember to write the decimal point and dollar sign in each answer.

1. $\begin{array}{r} \$5.30 \\ \times\ \ 3 \\ \hline \end{array}$ $\qquad$ $\begin{array}{r} \$4.20 \\ \times\ \ 4 \\ \hline \end{array}$ $\qquad$ $\begin{array}{r} \$9.32 \\ \times\ \ 3 \\ \hline \end{array}$ $\qquad$ $\begin{array}{r} \$21.11 \\ \times\ \ \ 8 \\ \hline \end{array}$ $\qquad$ $\begin{array}{r} \$6.24 \\ \times\ \ 2 \\ \hline \end{array}$ $\qquad$ $\begin{array}{r} \$7.20 \\ \times\ \ 4 \\ \hline \end{array}$

In the following problems, multiply by using the rules given on the previous page.

2. $\begin{array}{r} \$5.85 \\ \times\,100 \\ \hline \end{array}$ $\qquad$ $\begin{array}{r} \$7.36 \\ \times\,100 \\ \hline \end{array}$ $\qquad$ $\begin{array}{r} \$7.58 \\ \times\ \,10 \\ \hline \end{array}$ $\qquad$ $\begin{array}{r} \$6.50 \\ \times\,100 \\ \hline \end{array}$ $\qquad$ $\begin{array}{r} \$32.69 \\ \times\ \ \,10 \\ \hline \end{array}$ $\qquad$ $\begin{array}{r} \$7.49 \\ \times\,1,000 \\ \hline \end{array}$

Use a dollar sign and decimal point to write each answer below.

3. Multiply 92¢ by 4 $\qquad$ Find the product of 74¢ times 100 $\qquad$ Multiply 53¢ by 3

Multiplication Word Problems

A multiplication word problem requires you to multiply to determine an answer. Often, you are given a part of something and are asked to find the total. Or you may be given the price of a single object and then be asked to find the total price of several like objects. Here are two examples:

Example 1: Loretta's car gets 21 miles to the gallon. How many *total* miles can she expect to drive on a full tank of 16 gallons?

Answer: 336 miles

$$\begin{array}{r} 21 \\ \times 16 \\ \hline 126 \\ 21 \\ \hline 336 \end{array}$$

Example 2: If a quart of milk sells for 51¢, what is the price *of* 3 quarts?

Answer: $1.53

$$\begin{array}{r} \$0.51 \\ \times\ \ 3 \\ \hline \$1.53 \end{array}$$

Multiplication word problems may also contain key words. In Example 1, the word *total* is a **multiplication key word**. In Example 2, the second word *of* is a multiplication key word.

It is not surprising that many addition key words such as *total* are also multiplication key words. After all, both multiplication and addition are used to combine like things. Below are common multiplication key words.

Multiplication Key Words	
altogether	times
in all	total
of	twice
multiply	whole
product	

Remember, key words are only a clue to the solution of a problem. **A more important factor is your understanding of what the problem means.** Be encouraged to look for key words, but don't expect to find them in every problem.

Solve each multiplication problem below. Underline each multiplication key word that you can identify. Not all problems contain key words.

1. John's Men Store sold 21 shirts during a "Saturday Only Sale." If the store made a profit of $4 per shirt, what total profit did John's make on these shirts?

2. At an average highway speed of 52 miles per hour, how far can Larry travel in 4 hours?

3. Sheila puts $42 a month in a savings account for her infant son. At this rate, how much can she save in all during 1 year's time? (1 year = 12 months.)

4. After the weekend sale, a quart of oil will cost twice as much as it does today. What is the regular price if today's sale price is 63¢?

5. Jamie uses 2 yards of material for each dress she makes. If she plans to make 13 dresses during the next month, how many yards of material will she need?

6. On his new diet, Jeremy plans to lose 75 pounds. If he loses 4 pounds each month, how much can Jeremy expect to lose in one year?

7. To find the area of a rectangle, you multiply the length by the width. What is the area of a rectangle (in square feet) that measures 11 feet long and 8 feet wide?

8. June works 40 hours a week as a painter. If she also works an average of 12 overtime hours each week, how many overtime hours does she average each 4-week pay period?

Multiplying and Carrying

When the product of two digits is 10 or more, you must **carry**. As in addition, you place the carried digit at the top of the next column to the left. Then multiply and add the carried digit to the product. The example below shows how to multiply and carry to the tens column.

Example: Multiply: $\begin{array}{r} 45 \\ \times\ 7 \end{array}$ Solution: $\begin{array}{r} \overset{3}{4}5 \\ \times\ 7 \\ \hline 315 \end{array}$

Step 1. Multiply the ones digit: $7 \times 5 = 35$
Think of 35 as 3 tens and 5 ones.
Place the 5 under the ones column.

Step 2. Carry the 3 tens to the top of the tens column.

Step 3. Multiply the tens digit: $7 \times 4 = 28$
Add the carried 3 to 28: $28 + 3 = 31$
Write 31 under the tens and hundreds columns.

$\begin{array}{r} \overset{3}{4}5 \\ \times\ 7 \\ \hline 315 \end{array}$ Carry 3 tens to top of tens column. Place 5 under ones column.

Remember: **Multiply the tens digit before adding the carried digit.**

Carrying to the Tens Column

Multiply. Complete the partially worked Skill Builders before doing the problems.

Skill Builders						
1. $\begin{array}{r}\overset{2}{1}4\\ \times\ 6\\ \hline 4\end{array}$	$\begin{array}{r}\overset{3}{1}9\\ \times\ 4\\ \hline 6\end{array}$	$\begin{array}{r}\overset{2}{7}4\\ \times\ 7\\ \hline 8\end{array}$	$\begin{array}{r}\overset{1}{5}2\\ \times\ 8\\ \hline 6\end{array}$	$\begin{array}{r}1\overset{1}{2}6\\ \times\ 3\\ \hline 8\end{array}$	$\begin{array}{r}2\overset{3}{1}6\\ \times\ 5\\ \hline 0\end{array}$	$\begin{array}{r}7\overset{2}{2}9\\ \times\ 3\\ \hline 7\end{array}$

2. $\begin{array}{r}17\\ \times\ 2\\ \hline\end{array}$	$\begin{array}{r}15\\ \times\ 5\\ \hline\end{array}$	$\begin{array}{r}24\\ \times\ 4\\ \hline\end{array}$	$\begin{array}{r}36\\ \times\ 3\\ \hline\end{array}$	$\begin{array}{r}47\\ \times\ 2\\ \hline\end{array}$	$\begin{array}{r}29\\ \times\ 3\\ \hline\end{array}$	$\begin{array}{r}38\\ \times\ 2\\ \hline\end{array}$
3. $\begin{array}{r}54\\ \times\ 7\\ \hline\end{array}$	$\begin{array}{r}37\\ \times\ 6\\ \hline\end{array}$	$\begin{array}{r}42\\ \times\ 5\\ \hline\end{array}$	$\begin{array}{r}75\\ \times\ 8\\ \hline\end{array}$	$\begin{array}{r}48\\ \times\ 9\\ \hline\end{array}$	$\begin{array}{r}39\\ \times\ 4\\ \hline\end{array}$	$\begin{array}{r}37\\ \times\ 3\\ \hline\end{array}$
4. $\begin{array}{r}137\\ \times\ 2\\ \hline\end{array}$	$\begin{array}{r}128\\ \times\ 3\\ \hline\end{array}$	$\begin{array}{r}215\\ \times\ 2\\ \hline\end{array}$	$\begin{array}{r}129\\ \times\ 3\\ \hline\end{array}$	$\begin{array}{r}617\\ \times\ 4\\ \hline\end{array}$	$\begin{array}{r}428\\ \times\ 3\\ \hline\end{array}$	$\begin{array}{r}719\\ \times\ 2\\ \hline\end{array}$

Being Careful with Zeros

Multiplying numbers that have one or more zeros can be tricky. But you won't have any trouble if you remember these two rules:

- **The product of any number times 0 is 0.**
- **Multiply the zero first, then add the carried digit.**

Example:

$$\begin{array}{r} \overset{2}{5}07 \\ \times\ 3 \\ \hline 1,521 \end{array}$$

Step 1. $3 \times 7 = 21$ Write the 1, carry the 2.

Step 2. $3 \times 0 = 0$ Add the carried 2 to the 0.
$0 + 2 = 2$
Write the 2 in the tens place.

Step 3. $3 \times 5 = 15$ Write the 15.

Skill Builders

5.

$\overset{2}{2}08$	$\overset{2}{1}04$	$\overset{8}{1}09$	$\overset{4}{4}06$	$\overset{2}{8}07$	$\overset{3}{3}05$
$\times\ 3$	$\times\ 7$	$\times\ 9$	$\times\ 8$	$\times\ 4$	$\times\ 6$
4	8	1	8	8	0

6.

107	206	307	706	803	609
$\times\ 5$	$\times\ 4$	$\times\ 3$	$\times\ 7$	$\times\ 9$	$\times\ 6$

Carrying to the Hundreds Column

When the product of the tens digit is more than 10, you must carry to the hundreds column.

Example:

$$\begin{array}{r} \overset{1}{2}41 \\ \times\ 4 \\ \hline 964 \end{array}$$

Skill Builders

7.

$\overset{1}{1}62$	$\overset{2}{2}81$	270	$\overset{5}{2}91$	$\overset{4}{5}61$	740
$\times\ 2$	$\times\ 3$	$\times\ 3$	$\times\ 6$	$\times\ 7$	$\times\ 8$
24	43	0	46	27	0

8.

271	152	370	192	251	284
$\times\ 3$	$\times\ 4$	$\times\ 3$	$\times\ 4$	$\times\ 3$	$\times\ 2$

9.

482	571	620	461	580	281
$\times\ 4$	$\times\ 2$	$\times\ 9$	$\times\ 7$	$\times\ 6$	$\times\ 5$

Carrying to the Thousands Column

When the product of the hundreds digit is more than 10, you must carry to the thousands column.

Example:

$$\begin{array}{r} 1,\boxed{5}10 \\ \times \quad \boxed{5} \\ \hline 7,\boxed{5}50 \end{array}$$

Skill Builders					
10. $\overset{1}{3,310}$	$\overset{1}{2,423}$	$2,702$	$1,612$	$\overset{2}{5,723}$	$\overset{3}{6,821}$
$\times \quad 6$	$\times \quad 3$	$\times \quad 3$	$\times \quad 4$	$\times \quad 3$	$\times \quad 4$
860	269	106	448	169	284

11.	$1,412$	$2,621$	$1,813$	$2,910$	$3,611$	$1,713$
	$\times \quad 4$	$\times \quad 3$	$\times \quad 2$	$\times \quad 3$	$\times \quad 2$	$\times \quad 3$

12.	$5,721$	$4,632$	$8,210$	$7,723$	$9,610$	$7,924$
	$\times \quad 4$	$\times \quad 3$	$\times \quad 8$	$\times \quad 3$	$\times \quad 7$	$\times \quad 2$

Carrying to Two or More Columns

The following example shows how to carry to two or more columns.

Example: Multiply: $\begin{array}{r} 256 \\ \times \ 7 \\ \hline \end{array}$ Solution: $\begin{array}{r} \overset{3\ 4}{256} \\ \times \ 7 \\ \hline 1,792 \end{array}$

Step 1. Multiply the ones digit: $7 \times 6 = 42$
Write the 2 under the ones column.
Carry the 4 to the tens column.

$$\begin{array}{r} 2\,5\,\boxed{6}^{\boxed{4}} \\ \times \quad \boxed{7} \\ \hline \boxed{2} \end{array}$$

Step 2. Multiply the tens digit: $7 \times 5 = 35$
Add the carried 4: $35 + 4 = 39$
Write 9 under the tens column.
Carry the 3 to the hundreds column.

$$\begin{array}{r} \overset{\boxed{3}\ \boxed{4}}{2\,\boxed{5}\,6} \\ \times \quad \boxed{7} \\ \hline 17\,\boxed{9}\,2 \end{array}$$

Step 3. Multiply the hundreds digit: $7 \times 2 = 14$
Add the carried 3: $14 + 3 = 17$
Write 17 under the hundreds and thousands columns.

Skill Builders					
13. $\overset{2\ 2}{145}$	$\overset{1\ 2}{239}$	$\overset{3}{178}$	$\overset{2\ 2}{267}$	$\overset{7\ 4}{585}$	$\overset{4}{847}$
$\times \quad 5$	$\times \quad 3$	$\times \quad 4$	$\times \quad 4$	$\times \quad 9$	$\times \quad 7$
25	17	2	68	65	9

14.

167	148	262	376	192	173
× 6	× 5	× 7	× 2	× 5	× 6

15.

297	468	573	548	843	946
× 4	× 3	× 7	× 5	× 6	× 8

16.

$\overset{2\ 3\ 1}{1,453}$	$\overset{3\ 2}{1,275}$	$\overset{2}{1,184}$	$\overset{2\ 3}{5,759}$	$\overset{3\ 2}{3,643}$	$\overset{3}{4,687}$
× 6	× 5	× 7	× 4	× 8	× 5
718	75	8	036	44	5

17.

1245	2327	1265	2477	3586	1284
× 7	× 4	× 8	× 3	× 2	× 8

18.

3,576	5,375	2,486	4,586	6,478	3,580
× 5	× 4	× 7	× 9	× 3	× 5

Carrying: Mixed Practice

19.

273	38	1,254	307	345	209
× 3	× 2	× 6	× 5	× 7	× 8

20.

2,476	486	306	57	258	3,587
× 7	× 9	× 7	× 2	× 8	× 6

Carrying with Larger Numbers

When multiplying larger numbers, carrying may be needed to compute each partial product. The example below shows how to do this.

Example: Multiply: 247   Solution: 247
                  × 38   × 38

$$\begin{array}{r} 247 \\ \times\ 38 \\ \hline 1\,976 \\ 7\,41 \\ \hline 9{,}386 \end{array}$$

Step 1. Multiply 247 by 8: (3 and 5 are the carried digits)

$$\begin{array}{r} \overset{3\ 5}{247} \\ \times\ 38 \\ \hline 1976 \end{array}$$

NOTE: Each partial product has its own carried digits. It is a good idea to write carried digits lightly in pencil. Then you can erase them after you compute each partial product.

Step 2. Multiply 247 by 3: (1 and 2 are the carried digits)

$$\begin{array}{r} \overset{1\ 2}{247} \\ \times\ 38 \\ \hline 1976 \\ 741 \end{array}$$

Many people prefer just to remember the carried digits and not write them at all.

Step 3. Add the partial products.

Answer: **9,386**

Multiply. Complete each partially completed Skill Builders.

Skill Builders					
1.					

$$\begin{array}{r} 86 \\ \times 62 \\ \hline 172 \\ 516 \end{array} \qquad \begin{array}{r} 52 \\ \times 47 \\ \hline 364 \\ 8 \end{array} \qquad \begin{array}{r} 95 \\ \times 82 \\ \hline 190 \end{array} \qquad \begin{array}{r} 753 \\ \times\ 38 \\ \hline 6024 \\ 2259 \end{array} \qquad \begin{array}{r} 846 \\ \times\ 29 \\ \hline 7614 \\ 92 \end{array} \qquad \begin{array}{r} 937 \\ \times\ 43 \\ \hline 2\,811 \end{array}$$

2.

$$\begin{array}{r} 64 \\ \times 37 \\ \hline \end{array} \qquad \begin{array}{r} 83 \\ \times 54 \\ \hline \end{array} \qquad \begin{array}{r} 49 \\ \times 28 \\ \hline \end{array} \qquad \begin{array}{r} 79 \\ \times 48 \\ \hline \end{array} \qquad \begin{array}{r} 53 \\ \times 29 \\ \hline \end{array} \qquad \begin{array}{r} 75 \\ \times 64 \\ \hline \end{array}$$

3.

$$\begin{array}{r} 74 \\ \times 28 \\ \hline \end{array} \qquad \begin{array}{r} 58 \\ \times 33 \\ \hline \end{array} \qquad \begin{array}{r} 84 \\ \times 67 \\ \hline \end{array} \qquad \begin{array}{r} 27 \\ \times 19 \\ \hline \end{array} \qquad \begin{array}{r} 96 \\ \times 42 \\ \hline \end{array} \qquad \begin{array}{r} 45 \\ \times 36 \\ \hline \end{array}$$

4.

375	245	460	748	484	459
× 46	× 83	× 29	× 37	× 26	× 32

5.

683	365	565	295	730	625
× 46	× 27	× 67	× 38	× 29	× 35

Skill Builders

6.

534	368	364	947	374	684
× 263	× 356	× 326	× 420	× 207	× 503
1 602	2 208	2 184	18 940	2 618	2 052
32 04	18 40		378 8	0	
106 8					

7.

468	254	735	473	442	846
× 163	× 135	× 246	× 138	× 150	× 108

8.

407	518	607	730	364	730
× 285	× 341	× 445	× 378	× 470	× 307

9.

587	635	946	253	845	398
× 448	× 335	× 834	× 148	× 770	× 390

10.

482	376	947	1,630	1,846	3,374
× 407	× 317	× 705	× 706	× 354	× 836

Multiplying Numbers in a Row

When numbers are written in a row, rewrite them in columns. To simplify the multiplication, write the larger number as the top number.

Example: Multiply: 31 × 146

With larger number on top	*With smaller number on top*
146	31
× 31	×146
146	186
4 38	1 24
4,526	3 1
	4,526

NOTE: Worked either way, the answer is the same. However, with the larger number on top, you have fewer partial products to add.

Write the following problems in columns and multiply. Remember, on some problems you can use a shortcut.

1. 113 × 32   43 × 1,000   37 × 419

2. 217 × 109   568 × 46   293 × 814

3. 100 × 766   1,721 × 240   347 × 583

4. 154 × 10   23 × 589   400 × 1,663

Carrying with Dollars and Cents

To carry with dollars and cents, you carry from one column to the next as if the decimal point weren't there.

Example 1. Multiply:
$$\begin{array}{r} \$6.45 \\ \times\ \ 12 \\ \hline 12\ 90 \\ 64\ 5\ \ \\ \hline \$77.40 \end{array}$$

Step 1. Multiply the numbers.
645 × 12 = 7740

Step 2. Place the decimal point and dollar sign in the answer.

Answer: $77.40

Example 2: Multiply 87¢ by 26.

Step 1. Write 87¢ as $0.87

Step 2. Multiply:
$$\begin{array}{r} \$0.87 \\ \times\ \ 26 \\ \hline 5\ 22 \\ 17\ 4\ \ \\ \hline 22\ 62 \end{array}$$

Step 3. Place the decimal point and dollar sign in the answer.

Answer: $22.62

Multiply.

1.

$5.87	$4.76	$2.59	$6.78	$3.80	$5.70
× 8	× 4	× 7	× 3	× 6	× 4

2.

$7.58	$6.86	$4.67	$2.49	$6.08	$7.40
× 14	× 17	× 18	× 25	× 35	× 20

3.

$25.50	$15.30	$48.35	$15.69	$54.80
× 125	× 165	× 250	× 380	× 365

4. 97¢ times 8   48¢ times 7   83¢ times 5

5. 73¢ times 12   68¢ times 24   92¢ times 32

Working with Approximations

As discussed in Chapter 1, an **approximation** is a number that is "about equal" to an exact answer. For example, the number 75,000 is an approximation for the number 74,839.

In most problems it is easier to approximate an answer than it is to find an exact answer. Thus, approximation is a very useful tool when a "close answer" is all you need. For example, you need a "close answer" when a question asks "approximately how many . . . ?"

To determine an approximate answer, replace the numbers in the problem with round numbers that are easy to work with. Then solve the problem using the round numbers. (For a review of round numbers, reread pages 12 and 13.)

Example: Find an approximate answer to the addition problem at right.

$$\begin{array}{r} 692 \\ 403 \\ +289 \end{array}$$

Step 1. Round each number to the nearest hundred.
692 rounds to 700
403 rounds to 400
289 rounds to 300

Approximate Answer	*Exact Answer*
700	692
400	403
+300	+289
1,400	1,384

Step 2. Replace the original numbers with the round numbers and add.

Answer: 1,400 is an approximate answer.

Find an approximate answer to each problem below by rounding as indicated. The first one has been done for you.

In problems 1–4, round to the nearest 10.

1. $\begin{array}{r} 79 \ \textit{80} \\ +52 \ \textit{+50} \\ \hline \textit{130} \end{array}$

2. $\begin{array}{r} 187 \\ -\ 98 \end{array}$

3. $\begin{array}{r} 2,792 \\ 1,311 \\ +\ \ 978 \end{array}$

4. $\begin{array}{r} 512 \\ \times 203 \end{array}$

In problems 5 and 6, round to the nearest 1,000.

5. $\begin{array}{r} 34,893 \\ -12,796 \end{array}$

6. $\begin{array}{r} 26,923 \\ 17,090 \\ +\ 8,914 \end{array}$

In problems 7 and 8, round the top number to the nearest 100 and the bottom number to the nearest 10.

7. $\begin{array}{r} 516 \\ \times\ 38 \end{array}$

8. $\begin{array}{r} 711 \\ \times\ 89 \end{array}$

When using approximation, you must use your "math common sense." In any given problem, no one will tell you what round numbers to use. And you'll find that rounding to different numbers gives you very different approximate answers.

Look at this example. The exact answer is shown on the left. Two approximate answers are also shown. In the first, each number is rounded to the nearest 10 before multiplying. In the second, each number is rounded to the nearest 100.

Example:	849	*Exact Answer*	*Approximate Answer* **(nearest 10)**	*Approximate Answer* **(nearest 100)**
	× 394			

849		850	800
× 394		× 390	× 400
3 396		76 500	320,000
76 41		255 0	
254 7		331,500	
334,506			

As the example shows, rounding to a lower place value (10) gives an approximate answer that is much closer to the exact answer. However, rounding to a higher place value (100) simplifies the multiplication.

Find an approximate answer to each problem below. Use your math sense to decide what round numbers to use. There is no single correct choice to make when approximating an answer.

9.	79	10.	288	11.	4,597	12.	3,248
	42		237		2,203		2,196
	+18		+ 94		+1,994		+ 356

13.	89	14.	501	15.	3,401	16.	5,392
	−42		−297		−2,599		−1,607

17.	61	18.	179	19.	1,489	20.	3,492
	×19		× 41		× 39		× 103

Using Approximation in Word Problems

Approximation is a very useful tool in many word problems. Used carefully, approximation enables you to quickly choose an operation and check solutions.

To find an approximate answer to a word problem, substitute a round number for one or more numbers that appear in the problem. Then, solve the problem using the round numbers.

Example: Helen bought 9 cases of oil on sale for $11.95 per case. How much did she pay for all 9 cases?

To find an approximate answer, substitute $12 for $11.95 and multiply:

$12 × 9 = $108

Approximate answer: $108

NOTE: We also could have substituted 10 for 9 and gotten an answer of $120. However, $108 is closer to the exact answer, $107.55.

Notice how the approximate answer $108 serves as a useful clue:

- First, the amount $108 "feels" about right. This tells us that multiplication is probably the correct operation to use.

- Second, you know the exact answer should be close to $108. An answer of $1,007.55, $207.55, or another number far from $108 couldn't possibly be right. For these answers, your clue tells you to double-check your math.

A good rule to follow when using approximation is this:

★ **Substitute only enough to make an approximate answer easy to compute.**

You don't need to substitute a round number for every number that appears in a problem.

Following are addition, subtraction, and multiplication problems. In each problem, substitute round numbers as indicated and determine an approximate answer as shown in problem 1 below. Then, using this number as a clue, choose the correct answer from the choices given.

1. Georgia paid for a $13.19 purchase with a twenty-dollar bill. How much change should she get back?

   a) $3.81
   b) $6.81
   c) $13.81

   Substitute $13 for $13.19.

   Approximate answer: $20 - $13 = $7

   Exact answer: b) $6.81

2. Jan makes a commission of $2.89 on each pair of shoes she sells. What total commission will she make on sales of 31 pairs of shoes?

   a) $68.79
   b) $89.59
   c) $98.69

   Substitute $3 for $2.89 and 30 for 31.

   Approximate answer:

   Exact answer:

3. During his regular shift Thursday, Al carried three loads of gravel. The first load weighed 3,119 pounds; the second, 4,089 pounds; and the third, 3,912 pounds. How many total pounds of gravel did Al carry on Thursday?

   a) 11,120
   b) 19,120
   c) 21,120

   Substitute 3,000 for 3,119, 4,000 for 4,089, and 4,000 for 3,912.

   Approximate answer:

   Exact answer:

4. At a price of $14.89 each, what will 21 shirts cost?

   a) $148.29
   b) $249.19
   c) $312.69

   Substitute $15 for $14.89 and 20 for 21.

   Approximate answer:

   Exact answer:

5. As a fast-food hamburger cook, Rebecca is able to cook 197 hamburgers an hour during a busy day. At this rate, how many hamburgers could she cook on a busy 8-hour shift?

   a) 1,406
   b) 1,516
   c) 1,576

   Substitute 200 for 197.

   Approximate answer:

   Exact answer:

6. Starting with a 71-inch-long board, Bobby cut off a 42-inch piece. What was the length of the piece that remained?

   a) 29 inches
   b) 34 inches
   c) 39 inches

   Substitute 70 for 71 and 40 for 42.

   Approximate answer:

   Exact answer:

Solving Word Problems

On the next two pages there are addition, subtraction, and multiplication problems for you to solve. In each, compute an approximate answer and then solve the problem exactly. Write an approximate answer on the first line and the exact answer on the second line. The first problem is completed as an example.

1. What is the total cost of 19 gallons of gas at a price of $1.29 per gallon?

   $26.00 $24.51
   approximate   exact

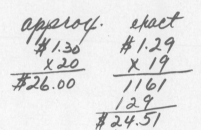

2. On Monday Alice drove 349 miles. On Tuesday she drove 497 miles. How far did Alice drive on these two days?

   _____   _____

3. Before "Big George" started his diet, he weighed 298 pounds. What is his weight now if he has lost 99 pounds?

   _____   _____

4. Norma's car gets 29 miles to the gallon while driving in the city. How many miles of city driving can she expect to do on a full tank of 15 gallons?

   _____   _____

5. How many calories are in the following meal?
   one hamburger: 403 calories
   one cola drink: 149 calories
   french fries: 296 calories

   _____   _____

6. When the half-cup size of Yummy Yogurt went on sale for 49¢, Mary bought all 31 on the shelf. What total amount did she pay for all this yogurt?

   _____   _____

Each problem on this page contains extra information. In each problem, underline the necessary information before solving. Then write an approximate answer on the first line and the exact answer on the second line.

7. When the price of lean hamburger was reduced from $1.89 to $1.49 per pound, Millie bought a 6-pound pack. How much did Millie pay for this meat?

   _____ _____

8. At a variety sale, Jack bought a lamp for $12.89, a chair for $3.49, and a box of magazines for $9.95. What total price did he pay for the lamp and chair?

   _____ _____

9. Mark earns $5.75 per hour as a sales clerk. He is also paid a commission of $6.98 for each suit he sells. How much did Mark earn in June from sales commissions for selling 51 suits?

   _____ _____

10. At 9:00 A.M. Mary Ann was 197 miles from Chicago. By 11:00 A.M. she had driven 98 miles closer. At that time, how far was she still from Chicago?

   _____ _____

11. The sticker on Jan's new car rates its mileage at 39 miles per gallon for highway driving and 27 miles per gallon for city driving. On a trip on the highway, how many miles can Jan drive if the tank contains 11 gallons of gas?

   _____ _____

12. From a 72-inch board, Shelley cut 3 pieces. These pieces measure 19 inches, 11 inches, and 29 inches. What is the total length of these 3 smaller boards?

   _____ _____

Multiplication Skills Review

Use these next two pages as a brief review of your multiplication skills. Work each problem as carefully as you can and check your answers with the answers on page 177. Correct any mistakes.

MULTIPLYING SMALL NUMBERS: Review pages 74 through 75.

1. 4   7   9   8   5   6 × 5 =
   ×2   ×1   ×6   ×1   ×8

MULTIPLYING LARGER NUMBERS, NO CARRYING: Review pages 76 through 78.

2. 32   60   143   34   231   423
   × 3   × 7   × 2   ×12   × 23   ×210

MULTIPLYING AND CARRYING: Review pages 84 through 89.

3. 15   46   423   309   381   384
   × 6   × 8   × 4   × 6   × 7   × 7

4. 47   86   50   482   705   375
   ×24   ×47   ×42   × 27   × 94   ×284

MULTIPLYING DOLLARS AND CENTS: Review pages 81 and 91.

5. $5.60 $20.14 $45.49 $4.58 $2.48 $12.40
   ×   6   ×   2   × 100   ×   9   ×   17   ×   34

MULTIPLYING NUMBERS WRITTEN IN A ROW: Review page 90.

6. 163 × 24   54 × 1,000   58 × 137   271 × 152

RECOGNIZING MULTIPLICATION WORD PHRASES: Review pages 82 through 83.

Write each word phrase in symbols and then solve.

7. thirty times nine

8. two hundred twenty-four multiplied by eighteen

9. the product of sixty and fifty

Write each amount in decimal form and then solve.

10. four dollars and six cents times 4

11. seventy-nine cents multiplied by 5

12. the product of one dollar and five cents and fifteen

SOLVING WORD PROBLEMS:

13. Expressed as dollars and cents, how much money is eighty-five quarters?

14. When gold is selling for $427 an ounce, what is the value of 1 pound of gold? (1 pound = 16 ounces)

15. Norma borrowed $5,000 from her credit union to buy a used car. To get the loan, she agreed to pay the credit union $134.14 each month for 48 straight months. What total amount of money will she pay the credit union over the 48-month period?

In problems 16 and 17, compute both an approximate answer and an exact answer. Write the approximate answer on the first line and the exact answer on the second line.

16. What is the cost of 19 gallons of gasoline at a price of $1.22 per gallon?

17. If Fran's car gets 41 miles to the gallon, how many miles can she expect to drive on 19 gallons?

_____ _____   _____ _____

5 Division Skills

Concepts in Division

To divide is to see how many times one number will "go into" a second number. For example, since there are four 2s in 8, 2 divided into 8 is 4.

In word problems, you often use division to find a part when you are given a total. Here are two examples:

- Divide to find the cost of one tire when four tires cost $248.

$$\begin{array}{r} \$62 \\ 4{\overline{\smash{)}248}} \end{array}$$

- Divide to find the distance driven in 1 hour when the distance driven in 3 hours is 150 miles.

$$\begin{array}{r} 50\text{ miles} \\ 3{\overline{\smash{)}150}} \end{array}$$

There are five other concepts that will help you improve your division skills.

1. In symbols, division is represented either by the division bracket ($\overline{)}$) or by the division sign ($\div$). The number being divided into is called the **dividend**. The number being divided by is called the **divisor**. The answer is called the **quotient**.

Parts of a Division Problem
Written with a division bracket *Written with a division sign*

$$\begin{array}{r} 9 \leftarrow \text{quotient} \\ 4{\overline{\smash{)}36}} \end{array}$$
divisor $\rightarrow$ $\quad$ $\llcorner$ dividend

$36 \div 4 = 9 \leftarrow$ quotient
$\llcorner$ divisor
$\llcorner$ dividend

2. Changing the order of divisor and dividend changes the value of the answer.

$$2{\overline{\smash{)}8}} \text{ is } not \text{ the same as } 8{\overline{\smash{)}2}}$$

3. The quotient of any number divided into 0 is 0.

$$5{\overline{\smash{)}0}} \qquad 0 \div 8 = 0$$

4. You can *never* divide by 0. Division by 0 has no meaning. However, you can divide by a number that ends in 0.

You can't divide by 0.

$0\overline{)8}$

You can divide by a number that ends in 0.

$10\overline{)50}^{5}$

5. In words, there are three ways to indicate division.

$8\overline{)56}$

Word Phrases

$\left\{ \begin{array}{l} 8 \underline{\text{divided into}} 56 \\ 56 \underline{\text{divided by}} 8 \\ \text{the } \underline{\text{quotient}} \text{ of } 56 \text{ divided} \\ \text{by } 8 \end{array} \right.$

Identify the divisor (number divided <u>by</u>), dividend (number divided <u>into</u>), and quotient (answer) of each solved problem below.

1. $9\overline{)72}^{8}$ divisor _____

   dividend _____

   quotient _____

2. $14 \div 7 = 2$ divisor _____

   dividend _____

   quotient _____

3. 24 divided by 4 is 6.

   divisor _____

   dividend _____

   quotient _____

4. 5 divided into 40 is 8.

   divisor _____

   dividend _____

   quotient _____

Use a division bracket to write each phrase below as a math problem. Remember that the number being divided up goes *inside* the bracket. Attach labels where given. You do not need to solve the problems.

5. two divided into sixteen hours

6. twenty-four dollars divided by six

7. the quotient of eighteen divided by three

8. 7 divided into 42 miles

9. 30 gallons divided by 10

10. the quotient of twenty-five divided by five.

Basic Division Facts

On these next two pages are the basic division facts you'll use in division problems. These facts come from the multiplication table.

To divide is really to "undo" multiplication. Look at these examples:

To solve:	You think:	In words:	Answer:
$8\overline{)24}$	$8 \times \underline{} = 24$	8 times <u>what number</u> equals 24?	3
$7\overline{)35}$	$7 \times \underline{} = 35$	7 times <u>what number</u> equals 35?	5
$56 \div 7$	$7 \times \underline{} = 56$	7 times <u>what number</u> equals 56?	8

Think about multiplication facts as you divide each problem below. Write both the division answer and the multiplication facts below. The first problem is done as an example.

1. $6\overline{)36}$ answer: 6

$(6 \times \underline{\;6\;} = 36)$

$3\overline{)12}$

$(3 \times \underline{} = 12)$

$5\overline{)40}$

$(5 \times \underline{} = 40)$

2. $8\overline{)56}$

$(8 \times \underline{} = 56)$

$2\overline{)4}$

$(2 \times \underline{} = 4)$

$9\overline{)36}$

$(9 \times \underline{} = 36)$

3. $7\overline{)14}$

$(7 \times \underline{} = 14)$

$2\overline{)16}$

$(2 \times \underline{} = 16)$

$3\overline{)27}$

$(3 \times \underline{} = 27)$

4. $8\overline{)72}$

$(8 \times \underline{} = 72)$

$4\overline{)12}$

$(4 \times \underline{} = 12)$

$5\overline{)20}$

$(5 \times \underline{} = 20)$

5. $54 \div 9 =$

$(9 \times \underline{} = 54)$

$18 \div 6 =$

$(6 \times \underline{} = 18)$

$81 \div 9 =$

$(9 \times \underline{} = 81)$

Divide as indicated. Write down multiplication facts if you find it helpful.

6. $2\overline{)10}$ $3\overline{)21}$ $7\overline{)49}$ $8\overline{)24}$ $2\overline{)18}$

7. $9\overline{)81}$ $4\overline{)16}$ $3\overline{)24}$ $9\overline{)45}$ $5\overline{)10}$

8. $3\overline{)9}$ $7\overline{)28}$ $4\overline{)32}$ $2\overline{)12}$ $6\overline{)18}$

9. $8\overline{)48}$ $5\overline{)15}$ $3\overline{)27}$ $2\overline{)8}$ $8\overline{)16}$

10. $63 \div 7 =$ $48 \div 6 =$ $20 \div 4 =$ $64 \div 8 =$

11. $45 \div 9 =$ $6 \div 3 =$ $28 \div 4 =$ $54 \div 6 =$

Write each word phrase in symbols and then solve. When given, be sure to include a label as part of the answer.

12. forty-five dollars divided by nine

thirty-two divided by eight

the quotient of 72 coins divided by 8

13. thirty-six divided by four

the quotient of 64 divided by 8

forty-two inches divided by seven

14. the quotient of twenty divided by five

56 divided by 8

twelve hours divided by four

Dividing by One Digit

After mastering the basic division facts, you are ready to divide larger numbers. To divide by a one-digit number, divide the dividend one digit at a time. Write your answer from left to right.

To *check* a division problem, multiply the answer by the divisor. The product should equal the dividend.

Example: Divide: $\dfrac{34}{2)\,68}$ Check: $\begin{array}{r} 34 \\ \times\ 2 \\ \hline 68 \end{array}$ (answer) (divisor) ✔

     Step 1.   Divide 2 into 6: $6 \div 2 = 3$ $\dfrac{3}{2)\,68}$
                Write 3 above the 6.

     Step 2.   Divide 2 into 8: $8 \div 2 = 4$ $\dfrac{34}{2)\,68}$
                Write 4 above the 8.

Answer: 34

Divide. Remember to work from left to right. Check each answer on scratch paper.

Dividing into 2-Digit Numbers

1. $4)\overline{84}$ $3)\overline{39}$ $2)\overline{48}$ $4)\overline{48}$ $2)\overline{22}$ $3)\overline{63}$ $2)\overline{28}$

2. $3)\overline{33}$ $4)\overline{44}$ $5)\overline{55}$ $2)\overline{46}$ $3)\overline{93}$ $2)\overline{24}$ $8)\overline{88}$

Dividing into 3-Digit Numbers

3. $4)\overline{844}$ $2)\overline{428}$ $3)\overline{693}$ $2)\overline{222}$ $2)\overline{282}$ $4)\overline{484}$

4. $2)\overline{224}$ $4)\overline{848}$ $3)\overline{663}$ $3)\overline{969}$ $2)\overline{842}$ $5)\overline{555}$

Dividing into Larger Numbers

5. $4)\overline{8,448}$ $2)\overline{4,824}$ $3)\overline{9,639}$ $2)\overline{62,484}$ $3)\overline{96,693}$ $4)\overline{84,488}$

Dividing into Zero

In many division problems the dividend contains one or more zeros. When you divide into zero, you get zero as an answer. Because the zero is part of a number, you must write 0 as part of the answer.

A zero may appear either at the end or in the middle of a dividend.

Example 1: Divide: $4\overline{)80}$ → 20

Step 1. Divide 4 into 8:
$8 \div 4 = 2$
Write 2 above the 8.

Step 2. Divide 4 into 0:
$0 \div 4 = 0$
Write 0 above the 0.

Answer: 20

Check:
20
× 4
——
80 ✔

Example 2: Divide: $3\overline{)609}$ → 203

Step 1. Divide 3 into 6:
$6 \div 3 = 2$
Write 2 above the 6.

Step 2. Divide 3 into 0:
$0 \div 3 = 0$
Write 0 above the 0.

Step 3. Divide 3 into 9:
$9 \div 3 = 3$
Write 3 above the 9.

Answer: 203

Check:
203
× 3
——
609 ✔

NOTE: Notice that if you had not written the zeros, your answers would have been 2 and 23. These are obviously wrong.

Divide. Check each answer on scratch paper.

Zeros on the End

1. $3\overline{)60}$ $2\overline{)40}$ $4\overline{)80}$ $2\overline{)640}$ $3\overline{)960}$ $2\overline{)820}$

2. $4\overline{)800}$ $5\overline{)500}$ $3\overline{)600}$ $2\overline{)400}$ $3\overline{)900}$ $2\overline{)800}$

Zeros in the Middle

3. $4\overline{)804}$ $2\overline{)608}$ $3\overline{)906}$ $4\overline{)808}$ $5\overline{)505}$ $3\overline{)609}$

4. $3\overline{)9,003}$ $4\overline{)8,008}$ $2\overline{)8,002}$ $5\overline{)5,005}$ $3\overline{)9,006}$ $4\overline{)4,008}$

Dividing into a Smaller Digit

In some problems you can't divide into the first digit of a number. It is smaller than the divisor! In this case, you divide into the first two digits.

Example: Divide: $\dfrac{52}{4\,)\,208}$ Check: $\begin{array}{r} 52 \\ \times\ 4 \\ \hline 208 \end{array}$ ✔

Step 1. You can't divide 4 into 2.
Divide 4 into 20: $20 \div 4 = 5$ $4\,)\,\overline{208}$ (5 above the 0)
Write 5 above the 0.

Step 2. Divide 4 into 8: $8 \div 4 = 2$ $\dfrac{52}{4\,)\,208}$
Write 2 above the 8.

Answer: 52

NOTE: In the example, you do not place a 0 in the answer because you do not divide 4 into 0. You divide 4 into 20. This problem differs from the problems on the previous page. *As this example shows, a zero in the dividend does not always mean you place a zero in the answer.*

Divide. Check each answer on scratch paper.

1. $5\,)\,\overline{255}$ $3\,)\,\overline{159}$ $6\,)\,\overline{186}$ $4\,)\,\overline{168}$ $2\,)\,\overline{1,862}$ $4\,)\,\overline{2,848}$

2. $5\,)\,\overline{405}$ $6\,)\,\overline{306}$ $2\,)\,\overline{108}$ $5\,)\,\overline{205}$ $4\,)\,\overline{2,048}$ $5\,)\,\overline{1,055}$

Each of the following problems has one or two zeros in the dividend. Some answers will contain zeros and some won't. Divide carefully.

3. $4\,)\,\overline{208}$ $5\,)\,\overline{405}$ $3\,)\,\overline{150}$ $6\,)\,\overline{606}$ $2\,)\,\overline{104}$ $7\,)\,\overline{140}$

4. $5\,)\,\overline{100}$ $3\,)\,\overline{300}$ $4\,)\,\overline{804}$ $3\,)\,\overline{2,706}$ $2\,)\,\overline{1,060}$ $4\,)\,\overline{2,008}$

Using Zero as a Place Holder

In many problems, it is the second or third digit that is smaller than the divisor. In these problems, you place a 0 in the answer to hold a place. Then you divide into the next two digits.

Example: Divide: $\dfrac{206}{3\overline{)618}}$ Check: $\begin{array}{r} 206 \\ \times\ \ 3 \\ \hline 618 \end{array}$ ✔

   Step 1. Divide 3 into 6: $6 \div 3 = 2$ $\dfrac{2}{3\overline{)618}}$
              Write 2 above the 6.

   Step 2. You can't divide 3 into 1. $\dfrac{20}{3\overline{)618}}$
              Write 0 above the 1.

   Step 3. Divide 3 into 18: $\dfrac{206}{3\overline{)618}}$
              $18 \div 3 = 6$
              Write 6 above the 8.

Answer: 206

Divide and check your answers.

1. $4\overline{)816}$ $4\overline{)412}$ $2\overline{)618}$ $3\overline{)927}$ $5\overline{)510}$ $6\overline{)636}$

2. $3\overline{)6,912}$ $4\overline{)8,416}$ $2\overline{)2,416}$ $4\overline{)4,420}$ $7\overline{)7,728}$ $5\overline{)5,510}$

Mixed Practice

3. $4\overline{)484}$ $3\overline{)609}$ $2\overline{)120}$ $7\overline{)350}$ $6\overline{)612}$ $3\overline{)900}$

4. $6\overline{)240}$ $7\overline{)714}$ $5\overline{)255}$ $8\overline{)640}$ $3\overline{)906}$ $4\overline{)820}$

5. $3\overline{)6,039}$ $4\overline{)8,040}$ $7\overline{)1,407}$ $6\overline{)3,600}$ $2\overline{)1,860}$ $5\overline{)2,505}$

Remainders in Division

A **remainder** is a number that is left over. For example, if you divide 13 record albums among 4 people, each person gets 3 albums. Since 4 × 3 is 12, one album is left over. Dividing 13 by 4 gives an answer of 3 with a remainder of 1.

Remainders are common in division. You get a remainder each time a number does not divide evenly. The letter *r* is used to stand for *remainder*. In the example above, we would write 13 ÷ 4 = 3 r 1.

To check a division problem that has a remainder, *multiply* and then *add* the remainder.

Example: Divide:

$$\begin{array}{r} 2\ r\ 3 \\ 5\overline{)\,13} \\ -10 \\ \hline 3 \end{array}$$

Check: Multiply *then* Add

$$\begin{array}{r} 2 \\ \times 5 \\ \hline 10 \end{array} \qquad \begin{array}{r} 10 \\ +\ 3 \\ \hline 13\ \checkmark \end{array}$$

Step 1. 5 does not divide evenly into 13. But 5 does divide evenly into 10. This is as close as we can get to 13.

10 ÷ 5 = 2

Write 2 above the 3.

$$\begin{array}{r} 2 \\ 5\overline{)\,13} \end{array}$$

Step 2. Multiply: 2 × 5 = 10
Write 10 beneath 13. Subtract: 13 − 10 = 3

$$\begin{array}{r} 2 \\ 5\overline{)\,13} \\ -10 \\ \hline 3 \end{array}$$

Step 3. Write the remainder 3 as *r 3* next to the 2.

Answer: 2 r 3

Divide. Check each answer. Complete the row of partially worked Skill Builders.

Skill Builders

1.

$$\begin{array}{r} 1\ r \\ 6\overline{)\,9} \\ -6 \\ \hline 3 \end{array} \qquad \begin{array}{r} 1 \\ 7\overline{)\,8} \\ -7 \\ \hline 1 \end{array} \qquad \begin{array}{r} 2 \\ 5\overline{)\,14} \\ -10 \end{array} \qquad \begin{array}{r} 2 \\ 8\overline{)\,21} \end{array} \qquad \begin{array}{r} 7 \\ 4\overline{)\,31} \end{array} \qquad \begin{array}{r} 7 \\ 6\overline{)\,44} \end{array}$$

2. $4\overline{)\,7}$ $\qquad$ $5\overline{)\,9}$ $\qquad$ $3\overline{)\,4}$ $\qquad$ $2\overline{)\,5}$ $\qquad$ $4\overline{)\,9}$ $\qquad$ $6\overline{)\,8}$

3.   5)‾17   8)‾58   9)‾62   4)‾27   5)‾38   7)‾51

To solve a division problem correctly, remember these two important rules:

- The product of the whole number part of the answer *times* the divisor (the number to the left of the division bracket) should always be smaller than the dividend (the number being divided up).

- The remainder should always be smaller than the divisor.

Example: Divide: 6)‾27

Solved Incorrectly		*Solved Correctly*

a)
$$\begin{array}{r} 5 \\ 6\overline{)27} \\ -30 \end{array}$$

b)
$$\begin{array}{r} 3\ r\ 9 \\ 6\overline{)27} \\ -18 \\ \hline 9 \end{array}$$

c)
$$\begin{array}{r} 4\ r\ 3 \\ 6\overline{)27} \\ -24 \\ \hline 3 \end{array}$$

a)   The answer 5 is incorrect because the product of 5 × 6 is 30, a number larger than the dividend, 27.

b)   The answer 3 r 9 is also incorrect. This is because the remainder, 9, is larger than the divisor, 6.

c)   Trying 4, we get the correct answer. The product of 4 × 6 is 24, a number smaller than 27. Also, the remainder is 3, a number smaller than the divisor, 6.

Cross out each problem below that is solved incorrectly. Then, in the space below, correct each problem that you crossed out.

4.

$$\begin{array}{r} 3\ r\ 8 \\ 5\overline{)23} \\ -15 \\ \hline 8 \end{array} \qquad \begin{array}{r} 4\ r\ 2 \\ 4\overline{)18} \\ -16 \\ \hline 2 \end{array} \qquad \begin{array}{r} 4\ r\ 13 \\ 7\overline{)41} \\ -28 \\ \hline 13 \end{array} \qquad \begin{array}{r} 7 \\ 9\overline{)61} \\ -63 \end{array} \qquad \begin{array}{r} 7\ r\ 1 \\ 3\overline{)22} \\ -21 \\ \hline 1 \end{array}$$

5.

$$\begin{array}{r} 8 \\ 10\overline{)78} \\ -80 \end{array} \qquad \begin{array}{r} 9\ r\ 2 \\ 5\overline{)47} \\ -45 \\ \hline 2 \end{array} \qquad \begin{array}{r} 7\ r\ 9 \\ 6\overline{)51} \\ -42 \\ \hline 9 \end{array} \qquad \begin{array}{r} 8\ r\ 8 \\ 8\overline{)72} \\ -64 \\ \hline 8 \end{array} \qquad \begin{array}{r} 7 \\ 7\overline{)49} \\ -49 \\ \hline 0 \end{array}$$

Introducing Long Division

Long division is a four-step process that you use for most division problems. Long division is used when you have a remainder before you finish dividing all of the digits.

Example: Divide: $4\overline{)92}$

Step 1	Step 2	Step 3	Step 4
$\begin{array}{r} 2 \\ 4\overline{)9\,2} \end{array}$	$\begin{array}{r} 2 \\ 4\overline{)92} \\ 8 \end{array}$	$\begin{array}{r} 2 \\ 4\overline{)92} \\ -8 \\ \hline 1 \end{array}$	$\begin{array}{r} 2 \\ 4\overline{)92} \\ -8 \\ \hline 12 \end{array}$

Step 1. Divide 4 into 9. Write 2 over the 9.

Step 2. Multiply: $2 \times 4 = 8$. Write 8 under the 9.

Step 3. Subtract: $9 - 8 = 1$. Write the remainder 1 under the 8.

Step 4. Bring down the 2 and place it next to the remainder 1.

Now repeat the four-step process. Start by dividing 4 into 12.

Step 1. Divide 4 into 12. Write the 3 over the 2.

Step 2. Multiply: $3 \times 4 = 12$. Write 12 under the 12.

Step 3. Subtract: $12 - 12 = 0$. Write 0 under the 12.

Step 4. There is no other digit to bring down, and there is no remainder. The problem is done.

$$\begin{array}{r} 2\,3 \\ 4\overline{)92} \\ -8 \\ \hline 1\,2 \\ -1\,2 \\ \hline 0 \end{array}$$

Answer: 23

Use long division to divide each problem below. Complete the rows of partially worked Skill Builders. Check each answer on scratch paper.

Skill Builders

1.

| $\begin{array}{r} 1 \\ 3\overline{)42} \\ -3 \\ \hline 12 \end{array}$ | $\begin{array}{r} 1 \\ 5\overline{)85} \\ -5 \\ \hline 3 \end{array}$ | $\begin{array}{r} 2 \\ 4\overline{)96} \\ -8 \\ \hline \end{array}$ | $\begin{array}{r} 2 \\ 6\overline{)138} \\ -12 \\ \hline 18 \end{array}$ | $\begin{array}{r} 1 \\ 8\overline{)136} \\ -8 \\ \hline 5 \end{array}$ | $\begin{array}{r} 8 \\ 5\overline{)435} \\ -40 \\ \hline \end{array}$ |

2. $2\overline{)36}$ $4\overline{)60}$ $6\overline{)78}$ $3\overline{)51}$ $8\overline{)96}$ $5\overline{)75}$

3. $4\overline{)136}$ $5\overline{)465}$ $3\overline{)237}$ $4\overline{)236}$ $7\overline{)434}$ $8\overline{)704}$

4. $5\overline{)425}$ $7\overline{)105}$ $9\overline{)198}$ $3\overline{)207}$ $4\overline{)300}$ $8\overline{)360}$

Answers with Remainders

As the example at right shows, many problems solved by long division also have a remainder as part of the answer. When you have "brought down" all of the numbers and have finished the division, you may still have a remainder. Solve and check your work.

Example:

```
     85 r 3
  4)343
   -32
     23
     20
      3
```

Skill Builders

5.

```
    18
  4)75
   -4
    35
    32
     3
```

```
    13
  5)67
   -5
    17
```

```
     1
  8)93
   -8
     1
```

```
    26
  6)157
   -12
     37
     36
      1
```

```
     5
  7)365
   -35
     15
```

```
     8
  3)268
   -24
```

---

6. $5\overline{)62}$ $7\overline{)89}$ $4\overline{)75}$ $8\overline{)95}$ $2\overline{)57}$ $3\overline{)83}$

7. $5\overline{)474}$ $6\overline{)523}$ $3\overline{)137}$ $7\overline{)692}$ $4\overline{)339}$ $2\overline{)175}$

111

Dividing Dollars and Cents

Divide dollars and cents by a whole number as if the decimal point weren't there. Then place the decimal point and dollar sign in the answer. Place the decimal point directly above its place in the dividend.

Example 1:

Divide: $6)\overline{\$4.14}$

Step 1. Use long division to divide the numbers.

$$6)\overline{\begin{array}{c}69\\ \$4.14\end{array}}$$
$$\underline{-3\,6}$$
$$54$$
$$\underline{54}$$
$$0$$

Step 2. Place the decimal point and dollar sign in the answer.

$$6)\overline{\begin{array}{c}\$0.69\\ \$4.14\end{array}}$$

Answer: $0.69

Example 2: What is the quotient of $0.72 divided by 9?

Step 1. Divide:
$$9)\overline{\begin{array}{c}8\\ \$0.72\end{array}}$$
$$\underline{-72}$$
$$0$$

Step 2. Place the dollar sign and decimal point in the answer.
Place the decimal point directly above its position in the dividend. Put a 0 between the decimal point and the 8.

$$9)\overline{\begin{array}{c}\$0.08\\ \$0.72\end{array}}$$

Answer: $0.08

Divide.

1. $4)\overline{\$1.52}$ $7)\overline{\$5.04}$ $6)\overline{\$2.04}$ $5)\overline{\$4.05}$ $3)\overline{\$1.65}$ $7)\overline{\$2.94}$

2. $3)\overline{\$0.24}$ $5)\overline{\$0.35}$ $6)\overline{\$0.60}$ $8)\overline{\$0.64}$ $5)\overline{\$0.45}$ $6)\overline{\$0.54}$

Use a dollar sign and decimal point to write each answer below.

3. What is the quotient of $7.56 divided by 7?   Divide $4.23 by 9.   What amount do you get by dividing $0.32 by 8?

Division Word Problems

A division word problem requires you to divide to determine an answer. Often, you are given a quantity and asked to find the size of a part. In other problems, you are given the price of several objects and asked to find the price of a single one. Here are two examples:

Example 1: On his trip to Oak Grove, Charlie drove 276 miles in 6 hours. On the *average*, how many miles did he drive *every* hour?

$$\begin{array}{r} 46 \\ 6\overline{)276} \\ -24 \\ \hline 36 \\ 36 \\ \hline 0 \end{array}$$

      Answer: 46 miles

Example 2: If yogurt is on sale for "Four for $1.84," how much does *each* container cost?

$$\begin{array}{r} \$0.46 \\ 4\overline{)\$1.84} \\ -1\,6 \\ \hline 24 \\ 24 \\ \hline 0 \end{array}$$

      Answer: $0.46

Division word problems also often contain key words in their questions. In Example 1, the words *average* and *every* are **division key words**. In Example 2, the word "each" is a division key word.

You can usually tell a division key word from a multiplication key word:

- A division key word refers to a single item. In fact, the most common division key words are *each* and *every*. Both of these words mean *one* or *single*. A multiplication key word refers to a total or to several items.

- Many division key words actually mean "divide." Example words are *split, share,* and *average*. Many multiplication key words actually mean "multiply." Example words are *product* and *times*.

By now you are probably becoming pretty good at recognizing key words. Look for example uses of some of the words in the following chart as you work the problems on the next two pages.

Division Key Words	
average	every
cut	one
divide(d)	share(d)
each	single
equal pieces	split

Solve each division problem below. Underline each division key word you can find. Each word problem may contain one or more key words.

1. When a deck of 52 cards is dealt to 4 people, how many cards does each person receive?

2. If spinach is selling for "Three bunches for $1.17," how much is a single bunch?

3. Six people won a lottery that had a value of $4,800,600. If they share the prize equally, what is each person's share?

4. During one 8-hour shift, Amy's Fast Food sold 176 hamburgers. On the average, how many hamburgers did Amy's sell during each hour?

5. If a large pizza is divided into 16 equal slices, how many pieces will each of 4 people get?

6. Anne worked 9 hours on Saturday on a part-time job. If she was paid $63.90 for this work, how much was she paid for every hour of work?

7. Louise wants to make 3 shelves out of a board that measures 147 inches long. About how long should she cut each of the equal pieces?

8. George and his 4 friends agreed to split a dinner bill of $20.50. Find how much George had to pay.

There are two special types of division word problems that often cause confusion.

In the first type, you are asked to find a whole number answer. However, when you divide, your answer contains a remainder. Here is an example:

Example 3: John is going to make 3-foot-long shelves out of a 20-foot-long board. How many shelves can he make?

$$\begin{array}{r} 6\ r\ 2 \\ 3\overline{)20} \\ -18 \\ \hline 2 \end{array}$$

As shown at right, $20 \div 3 = 6\ r\ 2$.

In this example, the question asks "how many shelves?"; this question requires a whole number answer. The answer is *6 shelves*. The leftover piece (2 feet) is less than a shelf and therefore is of no use.

In the second type of problem, you *divide* to find a *total*. In fact, the word *total* may be in the question. Yet, this is a division problem.

Example 4: If Virginia saves $5 a day in tips, what total number of days will it take her to save $135?

Answer: **27 days**

$$\begin{array}{r} 27 \\ \$5\overline{)\$135} \\ -10 \\ \hline 35 \\ 35 \\ \hline 0 \end{array}$$

NOTE: When both the divisor and the dividend contain a dollar sign, the answer doesn't.

Solve each division problem below.

9. If a dress takes two yards of material to make, how many dresses can be made from 9 yards?

10. Every week, Abe places $9 in a special savings account. How many total weeks will it take him to save enough money to buy the $126 portable tape recorder that he wants?

11. Amy packs and ships stereo records. She is able to pack 6 records in a single mailing box. How many boxes will she need to fill an order for 39 records?

12. Losing three pounds a month, how many months will it take Howard to lose a total of 51 pounds?

More About Long Division

Once you have learned the four steps of long division, you can divide any size number by repeating these steps.

Example: Divide: $7\overline{)952}$

1st Division: Divide 7 into 9.

Step 1. Divide: $9 \div 7 = 1$. Write 1 over the 9.
Step 2. Multiply: $1 \times 7 = 7$. Write 7 under the 9.
Step 3. Subtract: $9 - 7 = 2$. Write 2 under the 7.
Step 4. Bring down the 5 and place it next to the 2.

```
    1
7)9 52
 -7
  2 5
```

2nd Division: Divide 7 into 25.

Step 1. Divide: $25 \div 7 = 3$. Write 3 over the 5.

Step 2. Multiply: $3 \times 7 = 21$. Write 21 under the 25.

Step 3. Subtract: $25 - 21 = 4$. Write 4 under the 1.

Step 4. Bring down the 2 and place it next to the 4.

```
     1 3
7) 9 52
  -7
   2 5
  -2 1
    42
```

3rd Division: Divide 7 into 42.

Step 1. Divide: $42 \div 7 = 6$. Write 6 over the 2.

Step 2. Multiply: $6 \times 7 = 42$. Write 42 under the 42.

Step 3. Subtract: $42 - 42 = 0$. Write 0 under the 2.

Step 4. There is no other digit to bring down, and there is no remainder. The problem is done.

Answer: 136

```
    13 6
7)95 2
 -7
  25
 -21
   42
  -42
    0
```

Check your answer by multiplication:

$$\begin{array}{r} \overset{2\ 4}{136} \\ \times \quad 7 \\ \hline 952 \end{array}$$

Use long division to complete each problem below. Check each answer on scratch paper.

1.

$$
\begin{array}{r}
12 \\
8\overline{)976} \\
-8 \\
\hline
17 \\
16 \\
\hline
16
\end{array}
\qquad
\begin{array}{r}
1 \\
6\overline{)834} \\
-6 \\
\hline
23
\end{array}
\qquad
\begin{array}{r}
2 \\
3\overline{)744} \\
-6 \\
\hline
1
\end{array}
\qquad
\begin{array}{r}
26 \\
5\overline{)1,325} \\
-10 \\
\hline
32 \\
30 \\
\hline
2
\end{array}
\qquad
\begin{array}{r}
9 \\
9\overline{)8,334} \\
-81 \\
\hline
23
\end{array}
$$

2. $6\overline{)858}$ $3\overline{)822}$ $2\overline{)536}$ $8\overline{)936}$ $4\overline{)912}$

3. $3\overline{)2,214}$ $5\overline{)1,620}$ $2\overline{)1,738}$ $6\overline{)1,176}$ $7\overline{)4,438}$

Each problem below has a remainder in the answer.

4.

$$
\begin{array}{r}
163 \text{ r} \\
4\overline{)653} \\
-4 \\
\hline
25 \\
24 \\
\hline
13 \\
12 \\
\hline
1
\end{array}
\qquad
\begin{array}{r}
14 \\
3\overline{)440} \\
-3 \\
\hline
14 \\
12 \\
\hline
20
\end{array}
\qquad
\begin{array}{r}
1 \\
7\overline{)926} \\
-7 \\
\hline
22
\end{array}
\qquad
\begin{array}{r}
14 \\
8\overline{)1,146} \\
-8 \\
\hline
34 \\
32 \\
\hline
26
\end{array}
\qquad
\begin{array}{r}
4 \\
6\overline{)2,511} \\
-24 \\
\hline
11
\end{array}
$$

5. $6\overline{)715}$ $3\overline{)827}$ $5\overline{)928}$ $5\overline{)2,314}$ $7\overline{)3,726}$

Dividing by Two Digits

Dividing by a two-digit number is similar to dividing by a one-digit number. However, there is one difference. The first step is to *guess* what the answer will be! A good guess can save a lot of time and work.

Example 1:   Divide: $13\overline{)39}$

The multiplication facts don't tell you what $39 \div 13$ is. This is because these facts go only as high as number 9.
To find the above quotient, you make a good guess and then multiply to see if you're correct.

For example, try 3:
$$\begin{array}{r} 13 \\ \times\ 3 \\ \hline 39\ \end{array} \checkmark \qquad \begin{array}{r} 3 \\ 13\overline{)39} \\ -39 \\ \hline 0 \end{array}$$

3 is correct, and there is no remainder.

Answer: **3**

NOTE: The number 3 was chosen as a guess because the first number of the divisor (1) divides into the first number of the dividend (3) 3 times. This way of choosing a guess is helpful, but, as the next example shows, it doesn't always give the correct answer.

Example 2:   Divide: $16\overline{)67}$

Guess a quotient. Ask yourself, "About how many times does 16 go into 67?" Try several numbers.

Try 6:
$$\begin{array}{r} 16 \\ \times\ 6 \\ \hline 96 \end{array}$$
Try 5:
$$\begin{array}{r} 16 \\ \times\ 5 \\ \hline 80 \end{array}$$
Try 4:
$$\begin{array}{r} 16 \\ \times\ 4 \\ \hline 64\ \end{array} \checkmark \qquad \begin{array}{r} 4\ r\ 3 \\ 16\overline{)67} \\ -64 \\ \hline 3 \end{array}$$

4 is correct, and there is a remainder of 3.

Answer: **4 r 3**

NOTE: Guesses 6 and 5 are both too large. For trickier division problems such as this one, practice will improve your ability to guess correctly.

Divide. There are no remainders in rows 1 through 3. Check each answer on scratch paper.

Dividing into a Two-Digit Number

1. $13\overline{)26}$ $20\overline{)40}$ $14\overline{)28}$ $11\overline{)44}$ $12\overline{)24}$

2. $12\overline{)48}$ $22\overline{)66}$ $30\overline{)90}$ $11\overline{)55}$ $21\overline{)63}$

3. $14\overline{)70}$ $12\overline{)60}$ $16\overline{)64}$ $26\overline{)78}$ $13\overline{)52}$

Rows 4 and 5 may contain remainders.

4. $15\overline{)60}$ $14\overline{)90}$ $17\overline{)86}$ $25\overline{)75}$ $17\overline{)51}$

5. $12\overline{)96}$ $13\overline{)65}$ $29\overline{)63}$ $16\overline{)96}$ $16\overline{)52}$

Dividing into a Three-Digit Number

When the first two digits of the dividend are smaller than the two digits of the divisor, you must divide into the first three digits. At the right the answer 7 is placed above the 8.

Example:

$$\begin{array}{r} 7 \\ 24\overline{)168} \\ -168 \\ \hline 0 \end{array}$$

6. $25\overline{)125}$ $14\overline{)112}$ $34\overline{)170}$ $22\overline{)154}$ $40\overline{)160}$

7. $17\overline{)138}$ $28\overline{)140}$ $50\overline{)159}$ $19\overline{)120}$ $21\overline{)147}$

8. $16\overline{)112}$ $25\overline{)179}$ $32\overline{)160}$ $18\overline{)150}$ $26\overline{)130}$

Dividing Larger Numbers

The four-step process of long division is also used when you divide larger numbers. In the following example, we must divide twice to complete the problem.

Example: $23\overline{)828}$

1st Division: Divide 23 into 82.

   Step 1.   Divide: $82 \div 23 = 3$. Write 3 over the 2.

   Step 2.   Multiply: $3 \times 23 = 69$. Write 69 under the 82.

   Step 3.   Subtract: $82 - 69 = 13$. Write 13 under the 69.

   Step 4.   Bring down the 8 and place it next to the 13.

$$
\begin{array}{r}
3 \\
23\overline{)82\,8} \\
-69 \\
\hline
13\,8
\end{array}
$$

2nd Division: Divide 23 into 138.

   Step 1.   Divide: $138 \div 23 = 6$. Write 6 over the 8.

   Step 2.   Multiply: $6 \times 23 = 138$. Write 138 under the 138.

   Step 3.   Subtract: $138 - 138 = 0$. Write 0 under the 8.

   Step 4.   There is no other digit to bring down, and there is no remainder. The problem is completed.

$$
\begin{array}{r}
3\,6 \\
23\overline{)82\,8} \\
-69 \\
\hline
13\,8 \\
-13\,8 \\
\hline
0
\end{array}
$$

Answer: 36

Complete each problem below. Use multiplication to check each answer on scratch paper.

Dividing By Two-Digit Numbers

Skill Builders

1.

$$
\begin{array}{r}
1 \\
26\overline{)442} \\
-26 \\
\hline
182
\end{array}
\qquad
\begin{array}{r}
3 \\
17\overline{)646} \\
-51 \\
\hline
136
\end{array}
\qquad
\begin{array}{r}
3 \\
21\overline{)735} \\
-63 \\
\hline
105
\end{array}
\qquad
\begin{array}{r}
3 \\
31\overline{)1,054} \\
-93 \\
\hline
124
\end{array}
\qquad
\begin{array}{r}
4 \\
34\overline{)1,428} \\
-1\,36 \\
\hline
68
\end{array}
$$

2. $16\overline{)432}$ $27\overline{)486}$ $15\overline{)855}$ $28\overline{)532}$ $38\overline{)798}$

3. $32\overline{)1{,}312}$ $\qquad$ $46\overline{)1{,}426}$ $\qquad$ $39\overline{)1{,}092}$ $\qquad$ $52\overline{)1{,}144}$ $\qquad$ $35\overline{)1{,}750}$

For larger dividends, you may need to divide more than twice to complete the problem. Look carefully at the example at right.

Example:

$$
\begin{array}{r}
128 \\
37\overline{)4{,}736} \\
-37 \\
\hline
1\,03 \\
74 \\
\hline
296 \\
296 \\
\hline
0
\end{array}
$$

Skill Builders

4.

$$
\begin{array}{r}
12 \\
18\overline{)2{,}250} \\
-1\,8 \\
\hline
45 \\
36 \\
\hline
90
\end{array}
$$

$$
\begin{array}{r}
12 \\
24\overline{)3{,}024} \\
-2\,4 \\
\hline
62 \\
48 \\
\hline
\end{array}
$$

$$
\begin{array}{r}
2 \\
35\overline{)8{,}190} \\
-7\,0 \\
\hline
1\,19
\end{array}
$$

$$
\begin{array}{r}
2 \\
42\overline{)10{,}122} \\
-8\,4 \\
\hline
1\,72
\end{array}
$$

---

5. $19\overline{)2{,}413}$ $\qquad$ $17\overline{)3{,}621}$ $\qquad$ $25\overline{)3{,}125}$ $\qquad$ $27\overline{)6{,}318}$

6. $28\overline{)6{,}020}$ $\qquad$ $31\overline{)7{,}626}$ $\qquad$ $43\overline{)13{,}803}$ $\qquad$ $52\overline{)11{,}128}$

Dividing by Three-Digit Numbers

Dividing by three-digit numbers is similar to dividing by two-digit numbers. The first step is to divide into the first 3 (or 4) digits of the dividend.

In the example at right, 237 divides into 2014, the first 4 digits of the dividend. Two divisions are needed to complete this problem.

Example:

$$
\begin{array}{r}
85 \\
237\overline{)20{,}145} \\
-18\ 96 \\
\hline
1\ 185 \\
1\ 185 \\
\hline
0
\end{array}
$$

Skill Builders

7.

$$
\begin{array}{r}
4 \\
127\overline{)508}
\end{array}
\qquad
\begin{array}{r}
5 \\
213\overline{)1{,}065}
\end{array}
\qquad
\begin{array}{r}
1 \\
321\overline{)5{,}778} \\
-3\ 21 \\
\hline
2\ 568
\end{array}
\qquad
\begin{array}{r}
9 \\
219\overline{)20{,}367} \\
-19\ 71 \\
\hline
657
\end{array}
$$

---

8. $147\overline{)882}$ $\qquad$ $231\overline{)693}$ $\qquad$ $118\overline{)708}$ $\qquad$ $248\overline{)992}$

9. $312\overline{)1{,}560}$ $\qquad$ $418\overline{)1{,}672}$ $\qquad$ $279\overline{)2{,}232}$ $\qquad$ $303\overline{)1{,}212}$

10. $235\overline{)3{,}995}$ $\qquad$ $342\overline{)7{,}182}$ $\qquad$ $217\overline{)5{,}208}$ $\qquad$ $314\overline{)4{,}082}$

11. $216\overline{)16{,}848}$ $\qquad$ $322\overline{)17{,}388}$ $\qquad$ $297\overline{)18{,}117}$ $\qquad$ $403\overline{)23{,}374}$

Dividing Numbers in a Row

When a division problem is written with a ÷ sign, rewrite the problem with a division bracket ($\overline{)\quad}$). Remember, the number to the left of the ÷ sign is the dividend and goes inside the bracket.

Example: Divide: 773 ÷ 24

Rewrite as:

$$
\begin{array}{r}
32\ \text{r}\ 5 \\
24\overline{)773} \\
-72 \\
\hline
53 \\
48 \\
\hline
5
\end{array}
$$

Answer: 32 r 5

Divide. Check each answer on scratch paper. Some of the problems will have a remainder as part of the answer.

1. 364 ÷ 16   150 ÷ 25   462 ÷ 33   268 ÷ 12

2. 391 ÷ 17   792 ÷ 36   1,026 ÷ 27   1,127 ÷ 31

3. 3,240 ÷ 24   5,274 ÷ 41   924 ÷ 231   847 ÷ 205

4. 7,136 ÷ 223   8,543 ÷ 165   24,566 ÷ 346

Using Division to Check Multiplication

Division is often used to check the answer to a multiplication problem. To make this check, divide either of the two numbers being multiplied into the product (the answer of the multiplication). The answer to this division should equal the second number being multiplied.

Example: Use division to check the answer to the following multiplication problem.

$$534 \times 82 = 43{,}788$$

To check, divide either 534 or 82 into 43,788.

Check:
$$\begin{array}{r} 534 \\ 82{\overline{)43{,}788}} \\ -41\,0 \\ \hline 2\,78 \\ 2\,46 \\ \hline 328 \\ 328 \\ \hline \end{array}$$

Use division to check each multiplication problem below. The steps of the problems are not shown. Correct any answer that you find is incorrect.

1.
$$\begin{array}{r} 9 \\ \times 8 \\ \hline 72 \end{array}$$
$$\begin{array}{r} 8 \\ 9{\overline{)72}} \end{array}$$

2.
$$\begin{array}{r} 65 \\ \times 8 \\ \hline 530 \end{array}$$

3.
$$\begin{array}{r} 53 \\ \times 26 \\ \hline 1{,}378 \end{array}$$

4.
$$\begin{array}{r} \$2.12 \\ \times \quad 9 \\ \hline \$21.08 \end{array}$$

5.
$$\begin{array}{r} 204 \\ \times 43 \\ \hline 8{,}772 \end{array}$$

6.
$$\begin{array}{r} 309 \\ \times 31 \\ \hline 12{,}360 \end{array}$$

7.
$$\begin{array}{r} \$5.82 \\ \times \quad 14 \\ \hline \$81.48 \end{array}$$

8.
$$\begin{array}{r} 736 \\ \times 119 \\ \hline 85{,}584 \end{array}$$

9.
$$\begin{array}{r} 821 \\ \times 200 \\ \hline 164{,}200 \end{array}$$

Deciding When to Multiply and When to Divide

When multiplication and division problems appear on the same page, many people get confused. They are not always sure when to multiply and when to divide. This is not surprising. Although key words may be helpful, they are not always present.

In both multiplication and division problems, you are given two numbers and asked to find a third number:

- In multiplication problems, you are given *parts* of a total.
  You multiply these parts to determine the total.

- In division problems, you are given the *total* and *one part*.
  You divide the total by the given part to determine the unknown part.

While working on a problem, you may find it helpful to write a **solution sentence**. *A solution sentence uses words to state a problem's solution.* Once you write a solution in words, it is easy to replace words with numbers and then to compute the answer. Solution sentences are especially helpful for difficult problems.

Example 1: Stacey bought 4 cartons of pop for the birthday party. If each carton contains 6 bottles, how many bottles did she buy?

The problem is asking for the *total number of bottles* that Stacey bought. You know this even though the word *total* is not mentioned.

Here is a solution sentence for this problem:

total = cartons <u>times</u> bottles in a carton

To find this total, replace words with numbers:

total = 4 × 6
              (cartons)   (bottles)

total = 24 bottles

Answer: 24 bottles

Example 2: During a weekend clothing sale, Lewis sold $364 worth of A-1 work shirts. If he sold 26 shirts, what was the price of each shirt?

In this problem you are asked to find the price of a single shirt. The price of one shirt is a *part* of $364, the total cost of all 26 shirts.

For a solution sentence we can write:

price = total cost <u>divided by</u> number of shirts

To find this price, replace words with numbers:

price = $364 ÷ 26
      (total   (number
      cost)   of shirts)

price = $14

Answer: $14

Each problem below is followed by a *solution sentence*. Complete each sentence by writing either the word *multiplied* or the word *divided* on the blank line. Then, using your solution sentence as a guide, compute the answer to the question.

1. The owner of a grocery store bought 150 loaves of bread for $0.79 per loaf. What price did the owner pay for all the bread?

    total cost of bread = cost per loaf _____ by the number of loaves bought

2. Saving $2 each day, how long will it take Hank to save $330?

    number of days = total amount to be saved _____ by the amount saved each day

3. At $1.65 per pound, how much will 9 pounds of beef cost?

    total cost of beef = cost per pound _____ by the number of pounds bought

4. During a weekend trip, Kim drove 161 miles. If she used 7 gallons of gas, how many miles did her car run on each gallon?

    number of miles per gallon = total miles driven _____ by the number of gallons used

5. Farmer Murphy's cows produce 168 gallons of milk each day. Remembering there are 4 quarts in one gallon, how many quarts of milk is this each day?

    total quarts each day = number of gallons each day _____ by the number of quarts per gallon

Even when you clearly understand what a question asks for, you may not be sure whether to multiply or to divide. And you may find yourself confused about writing a correct solution sentence. When this happens, don't be afraid to solve the problem in more than one way. *Usually only one answer will make sense, and that will be the correct one!*

To the right of each problem below, <u>circle the correct answer</u>. Answer *a* is computed by multiplication; answer *b* by division. Use only common sense to help you decide which answer to choose. DO NOT SOLVE ANY PROBLEM.

6. Three roommates split their monthly rent evenly. If the rent is $465 per month, how much is each person's share?

   a) multiplication: $1,395
   b) division: $155

7. At a price of $1.28 each, how much will 8 cans of dogfood cost?

   a) multiplication: $10.24
   b) division: $0.16

8. If Gregory plans to drive 2,800 miles in 4 days, how many miles should he plan to drive each day?

   a) multiplication: 11,200
   b) division: 700

9. To lose 1 pound a week while dieting, a person needs to reduce his or her weekly calorie intake by 3,500 calories. How many calories is this each day?

   a) multiplication: 24,500
   b) division: 500

10. Craig's car gets 21 miles to the gallon. How many miles can Craig expect to drive on 7 gallons of gas?

   a) multiplication: 147
   b) division: 3

11. A local bakery sells chocolate chip cookies for $0.96 per dozen. How much will 6 dozen cookies cost?

   a) multiplication: $5.76
   b) division: $0.16

12. Four salesmen sold a total of $26,968 worth of furniture during April. On the average, how much furniture did each salesman sell that month?

   a) multiplication: $107,872
   b) division: $6,742

13. Jerry's Restaurant sold 352 small Cokes on Saturday. If each small cup holds 8 ounces of pop, how many ounces of Coke did Jerry's sell that day?

   a) multiplication: 2,816
   b) division: 44

Solving Word Problems

Solve the multiplication and division problems on the next four pages. As a first step in each problem, you may find it helpful to write a solution sentence. Then choose necessary information and compute an answer.

After solving each problem, check to see that your answer makes sense. If it doesn't, check to see that you used the correct operation, and check your math. Work on each problem until your answer seems correct.

1. If Virginia can type 312 words in 4 minutes, how many words can she type in one minute?

2. At a recent sale, Martha bought a case of oil for $18.96. The case contained 24 one-quart cans of oil. What amount did each quart of oil cost Martha?

3. Out of Stan's monthly check of $913, he must pay $285 for rent and an average of $125 for utilities. How much total rent does Stan pay each year?

4. Over the past 36 months, Lois has slowly lost 144 pounds. Her weight dropped from 284 pounds down to 140 pounds. During this time, what has been Lois's average weight loss per month?

5. Sonny reads at the rate of 240 words per minute. He once read an entire 88,000-word novel in just over 6 hours! At this rate, how many minutes will it take him to read a short story that contains 8,640 words?

6. Laura looked at a new oak dining set. The table alone cost $449. The set of four chairs costs $356. Laura bought the table and two of the four chairs. What price did Laura pay for each of the chairs?

7. Before José tuned up his car, it would get only 28 miles to the gallon. Now, after being tuned, it gets 37 miles to the gallon. How many miles can José hope to drive now on a full tank of 16 gallons?

8. Jenny is in charge of purchasing picnic supplies for Three Woods Lumber Company's annual picnic. Although 900 people work for the company, she expects that about 2500 cups will be needed. If a single carton holds 75 cups, how many cartons of cups should she get? She cannot buy partial cartons, so your answer must not have a remainder.

9. During a "TV Special" sale, Uptown Record Company shipped 1,238 boxes of records. If each box held 13 records, how many records were shipped in all?

10. "Quick Stop Oil" charges $15.99 for an oil and filter change. During a sale, they reduced the price to $13.49. On Monday of the sale, they serviced 47 cars. Find how much money they took in that day.

11. After buying a skirt for $32.99, Grace had $19.82 left. She decided to buy as many bottles of shampoo as she could. If each bottle of shampoo cost $2.00, how many bottles was she able to buy?

12. A 234-page book contains 72,072 words. On the average, how many words are written on a single page?

On the next two pages are short stories followed by questions. Each question requires its own necessary information. This information may be entirely in the story or it may be partly in the questions themselves. Choose this information carefully as you answer each question.

Questions 13 through 15 refer to the following story.

Carol applied for a job as a secretary with a publishing company. As part of the application process, she had to take a 6-minute typing test. During the 6 minutes, she typed 438 words. The boss was impressed, and Carol got the job.

Carol's starting salary was $935 per month. After 6 months she was given a raise. Her new salary was $1,156 per month.

13. How may words per minute did Carol type during the 6-minute typing test?

Information needed:

Solution:

14. How much did Carol earn during the first six months of her employment?

Information needed:

Solution:

15. After Carol received a raise, what was her new yearly salary?

Information needed:

Solution:

Questions 16 and 17 refer to the story below.

Tami borrowed $5,000 from Friendly Finance Company. To repay the loan, Tami must make 24 monthly payments of $237.71 each.

Before deciding on Friendly Finance, Tami also talked with Larry's Loan Company. Larry agreed to a $5,000 loan. But Tami would have to pay back a total of $5,764.80 divided into 24 equal monthly payments. After talking with Larry, Tami decided to get the loan from Friendly.

16. Over the 24-month loan period, what total payments will Tami make to Friendly Finance?

Information needed:

Solution:

17. If Tami had gotten the loan from Larry's Loan Company, what would her monthly payments have been?

Information needed:

Solution:

Questions 18 through 20 refer to the story below.

When they lived in Detroit, Mark and Ellen Johnson both had full-time jobs. Together they earned $540 each week.

After Mark lost his job, he and Ellen moved to Lansing. There Mark was able to get another job. His new job pays a yearly salary of $14,768. Ellen found part-time work that pays $4.75 per hour.

18. When Mark and Ellen lived in Detroit, what total income did they earn in one year?
(1 year = 52 weeks)

Information needed:

Solution:

19. In his new job in Lansing, how much does Mark earn each week?

Information needed:

Solution:

20. Working 18 hours per week, how much money can Ellen earn each week?

Information needed:

Solution:

Questions 21 through 24 refer to the information below.

Bill works as a shipping clerk in a food warehouse. His job is to pack cans of food into boxes in preparation for shipment. Last week, Bill was responsible for packing three fish products:

tuna fish — 128 cans per box
salmon — 48 cans per box
sardines — 144 cans per box

21. How many boxes does Bill need in order to ship 18,816 cans of tuna fish?

Information needed:

Solution:

22. How many boxes does he need in order to ship 4,368 cans of salmon?

Information needed:

Solution:

23. During the week, Bill shipped 98 boxes of sardines. How many cans of sardines did he ship?

Information needed:

Solution:

24. During this week, he also delivered a special order of 100 boxes of salmon. How many cans of salmon were in this special order?

Information needed:

Solution:

Division Skills Review

On the next two pages, review your division skills. Work each problem as carefully as you can and check your answers with the answers given on page 180. Correct any mistakes.

DIVIDING SMALL NUMBERS: Review pages 100 through 103.

1. $3\overline{)15}$ $4\overline{)36}$ $5\overline{)40}$ $6\overline{)54}$ $21 \div 7 =$ $28 \div 4 =$

DIVIDING LARGER NUMBERS, SHORT DIVISION: Review pages 104 through 107.

2. $4\overline{)84}$ $3\overline{)936}$ $2\overline{)680}$ $5\overline{)305}$ $4\overline{)4,080}$ $3\overline{)600}$

DIVIDING LARGER NUMBERS, LONG DIVISION: Review pages 110 through 111 and pages 116 through 122.

3. $4\overline{)25}$ $6\overline{)40}$ $5\overline{)75}$ $4\overline{)72}$ $7\overline{)377}$ $6\overline{)450}$

4. $6\overline{)275}$ $4\overline{)916}$ $12\overline{)252}$ $16\overline{)384}$ $35\overline{)875}$ $28\overline{)845}$

DIVIDING DOLLARS AND CENTS: Review page 112.

5. $5\overline{)\$1.50}$ $7\overline{)\$2.31}$ $3\overline{)\$0.96}$ $12\overline{)\$24.84}$ $25\overline{)\$37.50}$

DIVIDING NUMBERS WRITTEN IN A ROW: Review page 123.

6. $272 \div 4$ $385 \div 24$ $2,007 \div 9$ $42,845 \div 41$

RECOGNIZING DIVISION WORD PHRASES: Review pages 113 through 115.

Write each word phrase in symbols and then solve.

7. five divided into 60

8. the quotient of ninety divided by fifteen

9. thirty-six dollars divided by twelve

Write each amount in decimal form and then solve.

10. seventy-six cents divided by four

11. the quotient of four dollars and eighty cents divided by 8

12. fourteen divided into twenty-nine dollars and forty cents

SOLVING WORD PROBLEMS: Review pages 125 through 131.

13. When green onions are selling for "three bunches for $1.38," what price would you pay for a single bunch?

14. If Grace can type 65 words per minute, how many minutes will it take her to type a 2,990-word report?

15. Lynn makes and sells clay statues in her pottery shop. If she uses 6 pounds of clay for each statue, how many complete statues can she make with 44 pounds of clay?

16. For selling $8,493 worth of furniture in April, Harry was paid $942 in salary and $528 in commissions. Averaging a commission of $24 for each piece of furniture he sells, how many pieces did Harry sell during April?

17. For several months now, Betty has had her eye on a microwave oven that she'd like to buy. Its price has been $369.99. Last week when it went on sale for $289.58, Betty bought it. She paid $50.00 down and agreed to pay off the $239.58 balance in 6 equal monthly payments. If no interest is charged, how much will Betty pay each month?

6 Special Topics in Math

TOPIC 1: Introduction to Multi-Step Word Problems

Until now, you have worked with **one-step** word problems. A one-step problem can be solved by performing a single operation. Only one addition, subtraction, multiplication, or division is needed to compute the answer.

The solution of a **multi-step** word problem requires two or more operations. This type of problem is most easily solved by breaking it down into two or more one-step problems. Each step is then solved by a single operation.

To break down a multi-step problem into simpler one-step problems, it is best to begin by writing a **solution sentence**.

To write a solution sentence, use brief phrases and numbers to state a problem's solution as simply as possible.

Example 1: Raphael planned to spend $25.00 at Gene's Men Store. After buying a shirt for $14.45 and a tie for $1.95, how much money did Raphael have left?

To solve this problem, let's first write a solution sentence:

amount left = $25 **minus** cost of shirt and tie together

<u>missing information</u>

The total cost of shirt and tie is not given in the problem as a single number. Because it isn't, this amount is called **missing information**. Computing missing information is the first step in finding a problem's solution.

In example 1, the missing information is easily computed:

missing information = cost of shirt and tie together

$$= \$14.45 \ + \ \$1.95$$

$$= \mathbf{\$16.40}$$

Now the solution sentence can be written with numbers, and the answer can be computed by subtraction:

amount left = $\$25.00 \ - \ \$16.40$

$$= \mathbf{\$8.60}$$

Answer: $8.60

NOTE: Finding the missing information is a one-step problem that has its own *necessary information*. In example 1, to compute the cost of shirt and tie together, you identify the necessary information as $14.45 and $1.95. Then you add to find this total.

Example 2: Lynn found a great buy on school clothes at a Fall Clearance Sale. She bought 6 sweaters for $7.99 each. She also bought 5 skirts for $11.75 each. How much did Lynn pay for all these clothes?

A solution sentence can be written as follows:

cost of clothes = cost of sweaters *plus* cost of skirts

In this example, there are two items of *missing information*. Each has its own *necessary information*.

missing information	necessary information
total cost of sweaters	= 6 sweaters at $7.99
total cost of skirts	= 5 skirts at $11.75

Each value is computed by multiplication:

total cost of sweaters = $\$7.99 \times 6$

$$= \$47.94$$

total cost of skirts = $\$11.75 \times 5$

$$= \$58.75$$

Now we can place numbers in the solution sentence and compute the answer by addition:

total cost of clothes = $\$47.94 \ + \ \$58.75$

$$= \$106.69$$

Answer: $106.69

Solve each of the following problems. As a guide to each problem, we have written a solution sentence and two solution steps. In step 1, you compute the value of the missing information. In step 2, you complete the problem. The first one has been done for you.

1. At HI LIFE GROCERY, Jason bought hamburger for $3.58, a gallon of milk for $1.89, and a box of cereal for $2.79. If he paid the clerk with a twenty-dollar bill, how much change did he receive?

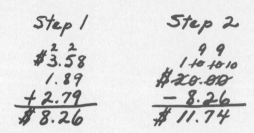

Solution sentence: change = $20 *minus* total cost of groceries

STEP 1. *Add* to find the total cost of groceries.

STEP 2. *Subtract* the total cost of groceries from $20 to determine Jason's change.

2. Each day Monday through Friday, Frieda delivers a newspaper to each of her 121 customers. On last weekend she delivered a total of 257 papers. What total number of papers did Frieda deliver last week?

Solution sentence: total papers = Monday-Friday total *plus* 257

STEP 1. *Multiply* the daily total by 5 to find the Monday-Friday total.

STEP 2. *Add* the Monday-Friday total to 257 to determine the total papers delivered.

3. Flora baked cookies for her first grade class. From the 84 cookies she baked, she kept 12 at home for her family. She took the rest to school. If there are 24 kids in her class, how many cookies can each child have at snack time?

Solution sentence: cookies per child = cookies for class *divided by* 24

STEP 1. *Subtract* to find the number of cookies for the class.

STEP 2. *Divide* the cookies for the class by 24 to determine the cookies for each child.

Solve each problem below. As a first step, complete the solution steps by writing either *add, subtract, multiply,* or *divide* on each blank line.

4. Five friends agreed to split the cost of dinner evenly. Together they had a pizza for $12.75, drinks for $4.30, and salads for $9.45. What was each person's share of the bill?

   Solution sentence: each share = total cost *divided by* 5

   STEP 1. _____ to find the total cost of dinner.

   STEP 2. _____ the cost of dinner by 5 to determine each person's share.

5. For her church, Shelley made 23 gift packages to give to children following Easter services. Each package contained 79¢ worth of candy, a $1.98 toy, and a 35¢ balloon. How much was the total cost of all 23 packages?

   Solution sentence: total cost = cost of 1 package *times* 23

   STEP 1. _____ to find the cost of 1 package.

   STEP 2. _____ the cost of 1 package by 23 to determine the total cost of all packages.

6. George bought 15 cases of oil to use in his car tune-up business. Each case contains 24 quarts. If it takes 5 quarts for each tune-up, how many tune-ups can George do before he needs more oil?

   Solution sentence: number of tune-ups = total quarts *divided by* 5

   STEP 1. _____ to find the total quarts of oil purchased.

   STEP 2. _____ total quarts purchased by 5 to determine the number of tune-ups that can be done.

Solve each problem below. Use the solution sentence written beneath each problem as a guide to your work.

7. Brenda, a plumber, fixed the sink at Mark's house. She charged Mark a total of $75.40. The cost of new parts included $28.50 for a new faucet, $13.65 for tubing, and $0.83 for washers. How much did Brenda charge for her labor?

   Solution sentence: labor cost = $75.40 *minus* total cost of parts

8. In Dick's pickup he can carry 8 boxes. He can also carry 6 more boxes in a small trailer attached to his pickup. How many boxes can Dick move in 8 trips if he uses both pickup and trailer?

   Solution sentence: total boxes moved = boxes carried per trip *times* 8.

9. Three clubs rented the Town Hall for a celebration dance. The hall rent was $85. The cost of the band was $285, and refreshments cost $98. If they split all costs evenly, what amount was each club's total share?

   Solution sentence: each share = total expenses *divided by* 3

10. As part of a sales idea, Bobbie gave away 55 balloons every day during the 30 days in April. At the end of this month, she still had 245 balloons left over. Knowing this, figure out how many total balloons Bobbie started out with.

   Solution sentence: total balloons = number of balloons given away *plus* 245

For each problem below, write a brief solution sentence to show how the problem is solved. Then solve the problem using your solution sentence as a guide.

11. Marie bought three dresses during the Spring Clearance Sale. One dress cost $34.95, one cost $27.50, and the other cost $19.95. If she paid for the dresses with a hundred-dollar check, how much change should she receive?

    Solution sentence:

12. For the first meeting of his First Aid Class, Bill bought 17 manuals. Each manual cost $6.49. He also bought a first aid kit for $17.50. Find the total cost of these class supplies.

    Solution sentence:

13. Working as a waitress Monday through Thursday, Lita serves about 95 people each day. On Saturday she serves about 165. During the 5 days that Lita works, about how many people does she serve?

    Solution sentence:

14. John and his two brothers split the cost of renting a small mobile home to take hunting. Daily rental cost is $27.84. If they kept the unit for 5 days, how much would be each brother's share of the rental cost?

    Solution sentence:

Below are three multi-step word problems for you to solve. Each contains extra information. In each problem, do the following three things:

1) Write a brief solution sentence.
2) Underline all necessary information.
3) Using your solution sentence as a guide, compute the answer.

15. Angie traded in her old Ford on a newer Toyota. The Ford had 119,724 miles on it. The Toyota showed only 32,696 miles. During the next twelve months, Angie drove the Toyota an average of 1,350 miles each month. At the end of the year, how many miles now showed on the Toyota's mileage indicator?

16. Heather bought a tube of toothpaste for $2.89, a hairbrush for $3.49, and bathroom tissue for 79¢. Although she had a ten-dollar bill in her purse, she decided to pay by check. For how much should she write the check if she wants to receive $5 back in change?

17. In Lewisville there are 26,854 registered voters. During the fall election, 14,725 people voted for Measure #1. Of those who voted, 6,539 were Republicans and 4,891 were Democrats. How many people who voted for Measure #1 were neither Republicans nor Democrats?

TOPIC 2: Introduction to Measurement

Measurement is an important part of daily life. Common uses include the measuring of length, weight, liquid measure, and time. A measurement is given as a number and a label. The label is called a **measurement unit**. For example, in the length 14 feet, *feet* is the measurement unit.

There are two types of measurement units in use in the United States. One is the familiar **English System**. The other is the **Metric System**. Commonly used units in each system are shown below.

UNITS OF LENGTH

ENGLISH SYSTEM		METRIC SYSTEM	
Unit	*Relation to Other Units*	*Unit*	*Relation to Other Units*
inch (in. or ″)	Smallest unit	millimeter (mm)	Smallest unit
foot (ft. or ′)	1 ft. = 12 in.	centimeter (cm)	1 cm = 10 mm
yard (yd.)	1 yd. = 36 in.	meter (m)	1 m = 1,000 mm
	1 yd. = 3 ft.		1 m = 100 cm
mile (mi.)	1 mi. = 5,280 ft.	kilometer (km)	1 km = 1,000 m
	1 mi. = 1,760 yd.		

1 meter ▬▬▬▬▬

1 yard ▬▬▬▬▬

(to scale)

1 kilometer ▬▬▬

1 mile ▬▬▬▬

(to scale)

1 cm ▬▬

1 inch ▬▬▬

UNITS OF WEIGHT

ENGLISH SYSTEM		METRIC SYSTEM	
Unit	*Relation to Other Units*	*Unit*	*Relation to Other Units*
ounce (oz.)	Smallest unit	milligram (mg)	Smallest unit
pound (lb.)	1 lb. = 16 oz.	gram (g)	1 g = 1,000 mg
ton (T.)	1 T. = 2,000 lb.	kilogram (kg)	1 kg = 1,000 g
		metric ton (t)	1 t = 1,000 kg

1 kilogram → ← 1 pound

A kilogram is a little heavier than 2 pounds.

UNITS OF LIQUID MEASURE

ENGLISH SYSTEM

Unit	Relation to Other Units
ounce (oz.)	Smallest unit
pint (pt.)	1 pt. = 16 oz.
quart (qt.)	1 qt. = 2 pt.
gallon (gal.)	1 gal. = 4 qt.

METRIC SYSTEM

Unit	Relation to Other Units
milliliter (ml)	Smallest unit
liter (L)	1 L = 1,000 ml
kiloliter (kl)	1 kl = 1,000 L

1 liter →   ← 1 quart

A liter is slightly larger than a quart.

UNITS OF TIME

Units of time are identical in both English and Metric Systems.

Unit	Relation to Other Units
second (sec.)	Smallest unit
minute (min.)	1 min. = 60 sec.
hour (hr.)	1 hr. = 60 min.

Unit	Relation to Other Units
day (da.)	1 da. = 24 hr.
week (wk.)	1 wk. = 7 da.
year (yr.)	1 yr. = 365 da.

To become more familiar with metric units, do the exercises below.

Fill in each blank with one of the choices within parentheses.

1. A meter is _____ than a yard. (longer, shorter)

2. A kilogram is _____ than a pound. (heavier, lighter)

3. A liter is _____ than a quart. (larger, smaller)

4. A speed of 50 kilometers per hour is _____ a speed of 50 miles per hour. (greater than, less than)

Answer each question below.

5. The smallest metric unit of length is the _____.

6. The smallest metric unit of weight is the _____.

7. The smallest metric unit of liquid measure is the _____.

Circle the larger in each pair of quantities.

8. 10 yards *or* 10 meters _____.

9. 5 pounds *or* 5 kilograms _____.

10. 3 quarts *or* 3 liters _____.

Finding the Number of Smaller Units in a Larger Unit

Sometimes you may need to find the number of smaller units in a larger unit, and that number may not be written in a table. The following example will show what we mean.

Example: How many pints are in a gallon?

From the table on page 141 we see that we need the following two relationships:

1 quart = 2 pints

1 gallon = 4 quarts

Although the number of pints in a gallon is not directly given, we can easily determine this number by multiplication:

pints in a gallon = pints in a quart *times* quarts in a gallon

$$= 2 \times 4$$

$$= 8$$

Answer: 1 gallon = 8 pints

As the example shows, the first step is to write down the relationships that connect the units you are interested in. Then multiply to find the answer.

Determine each number as indicated below.

1. the number of seconds in an hour

2. the number of inches in a mile

3. the number of centimeters in a kilometer

4. the number of grams in a metric ton

5. the number of liquid ounces in a quart

6. the number of millimeters in a kilometer

Changing from One Unit to Another

A quantity can be written in more than one unit. For example, you can write 1 yard as 3 feet or as 36 inches. To **convert** (change) from one unit to another, you use a **conversion factor**.

A conversion factor is a number that relates two different units. The tables of measurements on pages 140 and 141 list common units and their conversion factors.

Here are a few examples:

Units	Conversion Factor	Units	Conversion Factor
1 foot = 12 inches	12	1 pound = 16 ounces	16
1 yard = 3 feet	3	1 meter = 100 centimeters	100

To change from one unit to another, you use a conversion factor as follows:

To change larger units to smaller units, <u>multiply</u> by the conversion factor.

Example 1: Change 6 yards to feet (feet are smaller than yards).

Step 1. Identify the conversion factor that relates yards and feet. Since 1 yard = 3 feet, *the conversion factor is 3.*

Step 2. To change yards to feet, multiply by 3.

6 yards = 6 × 3 = 18 feet

Answer: 18 feet

To change smaller units to larger units, <u>divide</u> by the conversion factor.

Example 2: Change 64 ounces to pounds (pounds are larger than ounces).

Step 1. Identify the conversion factor that relates ounces and pounds. Since 1 pound = 16 ounces, *the conversion factor is 16.*

Step 2. To change ounces to pounds, divide by 16.

64 ounces = 64 ÷ 16 = 4 pounds

Answer: 4 pounds

Change each quantity to the unit indicated. As your first step in each problem, identify the correct conversion factor.

Changing Larger Units to Smaller Units: Multiply

1. 7 ft. = _____ in.   8 yd. = _____ ft.   12 cm = _____ mm

2. 3 lb. = _____ oz.   5 km = _____ m   2 kg = _____ g

3. 3 qt. = _____ pt.   2 L = _____ ml   4 min. = _____ sec.

Changing Smaller Units to Larger Units: Divide

4. 36 in. = _____ ft.   300 cm = _____ m   12 pt. = _____ qt.

5. 72 hr. = _____ da.   12 qt. = _____ gal.   120 sec. = _____ min.

6. 24 ft. = _____ yds.   4,000 lb. = _____ T.   80 oz. = _____ lb.

Many times, when smaller units are converted to larger units, there is a remainder.

As the example at the right shows, this remainder is simply written as the number of smaller units left over.

Example: Change 7 feet to yards.

Step 1.   Divide 7 feet by the conversion factor 3.

7 ÷ 3 = 2 r 1

Step 2.   Write the remainder as the number of feet left over.

7 feet = 2 yards 1 foot

Change each quantity to the unit indicated. Write each remainder as the number of smaller units left over.

7. 29 in. = _____ ft. _____ in.   14 ft. = _____ yds. _____ ft.

8. 75 min. = _____ hr. _____ min.   47 oz. = _____ lbs. _____ oz.

9. 240 cm = _____ m _____ cm   26 mm = _____ cm _____ mm

10. 2,500 g = _____ kg _____ g   3,400 ml = _____ L _____ ml

Simplifying Mixed Units

A quantity such as 6 ft. 4 in. is said to be in **mixed units** because it contains more than one unit (in this case, ft. and in.). When possible, you should simplify a mixed unit quantity by changing smaller units to larger units.

Example: Simplify 5 lb. 19 oz.

5 lb. 19 oz. can be simplified because 19 oz. is larger than 1 lb.

Step 1.   Change 19 oz. to lb. and oz. To do this, divide by the conversion factor 16 (1 lb. = 16 oz.).

$$19 \div 16 = 1 \text{ r } 3$$
$$= 1 \text{ lb. } 3 \text{ oz.}$$

Step 2.   Add the 1 lb. 3 oz. to the 5 lb.

$$\begin{array}{r} 5 \text{ lb.} \\ +1 \text{ lb. } 3 \text{ oz.} \\ \hline 6 \text{ lb. } 3 \text{ oz.} \end{array}$$

Answer   5 lb. 19 oz. = 6 lb. 3 oz.

Simplify each quantity below.

1.   5 ft. 31 in. = ____ ft. ____ in.   23 cm 14 mm = ____ cm ____ mm

2.   9 yd. 13 ft. = ____ yd. ____ ft.   2 lb. 33 oz. = ____ lb. ____ oz.

3.   13 m 350 cm = ____ m ____ cm   3 qt. 7 pt. = ____ qt. ____ pt.

4.   5 L 2,750 ml = ____ L ____ ml   4 min. 187 sec. = ____ min. ____ sec.

5.   7 ft. 44 in. = ____ ft. ____ in.   3 kg 2,300 g = ____ kg ____ g

Adding Measurement Units

To add two or more quantities, add each unit of measurement separately and then simplify the sum. An example is shown below for lengths in both the English and Metric Systems.

ENGLISH SYSTEM

Add: 8 ft.   7 in.
          +7 ft.   9 in.
          15 ft. 16 in.

Answer: 16 ft. 4 in.

(Since 16 in. = 1 ft. 4 in.)

Addition
Step 1. Add each column separately.
Step 2. Simplify the sum.

METRIC SYSTEM

Add: 6 cm   9 mm
          +5 cm   4 mm
          11 cm 13 mm

Answer: 12 cm 3 mm

(Since 13 mm = 1 cm 3mm)

Add the following quantities and simplify each sum.

1.   4 yd. 2 ft.
       +2 yd. 2 ft.

        5 yd. 1 ft.
       +3 yd. 2 ft.

        9 ft. 7 in.
       +7 ft. 8 in.

        3 mi. 1,500 yd.
       +2 mi   450 yd.

2.   7 cm 9 mm
       +5 cm 3 mm

        6 cm 5 mm
       +2 cm 8 mm

        25 m 35 cm
       +12 m 86 cm

        47 km 750 m
       +30 km 500 m

3.   5 lb.   7 oz.
       +2 lb. 14 oz.

        6 lb. 12 oz.
       +3 lb. 11 oz.

        1 lb. 14 oz.
       +   15 oz.

        3 T. 1,300 lb.
       +1 T.   675 lb.

4.   8 kg 900 g
       +5 kg 750 g

        3 kg 700 g
       +1 kg 535 g

        12 g 125 mg
       + 9 g 900 mg

        2 t 574 kg
       +1 t 875 kg

5.   5 qt. 1 pt.
       +2 qt. 1 pt.

        4 qt. 1 pt.
       +3 qt. 1 pt.

        3 pt. 9 oz.
       +1 pt. 8 oz.

        1 gal. 3 qt.
       +1 gal. 2 qt

6.   2 L 800 ml
       +1 L 750 ml

        3 L 750 ml
       +2 L 500 ml

        25 L 500 ml
       +   625 ml

        2 kl 900 L
       +1 kl 350 L

7.   3 hr. 35 min.
       +2 hr. 45 min.

        2 hr. 18 min.
       +1 hr. 50 min.

        7 min. 45 sec.
       +5 min. 19 sec.

        2 da. 18 hr.
       +1 da.   7 hr.

Solve each word problem below.

8. Karin made a long garden hose by connecting 3 shorter hoses together. The lengths of the shorter hoses were 5 yd. 2 ft., 4 yd. 1 ft., and 6 yd. 2 ft. What is the combined length of these three hoses?

9. On Saturday Ellen went on a 20 km hike. During the first hour, she walked 4 km 800 m. During the second hour, she walked 5 km 400 m, and during the third, 6 km 100 m. What total distance did she walk in 3 hours?

10. Ervin put a turkey in the oven at 25 minutes after 2:00. If he's supposed to cook it for 3 hours and 45 minutes, at what time should he turn off the oven?

11. A French punch recipe calls for 1 liter 500 milliliters of pineapple juice, 350 milliliters of grapefruit juice, and 2 metric cups of yogurt. (A metric cup equals 250 milliliters.) What is the combined volume of these three ingredients?

12. Sheila bought 3 steaks while shopping at the meat market. The large one weighed 2 lb. 7 oz. The smaller ones weighed 1 lb. 3 oz. and 13 oz. What is the total weight of steak that Sheila bought?

13. At Wong's Market, Yoshi bought 3 bags of rice, each weighing 10 kg, 750 g. What is the total weight of rice that Yoshi bought?

Subtracting Measurement Units

To subtract one quantity from another, subtract each unit of measurement separately, starting with the unit in the right-hand column. Borrowing from a larger unit is often necessary and is done in a way similar to borrowing in whole numbers. A borrowed larger unit is changed into smaller units. It is then added to the smaller units already present in the problem. An example of borrowing is shown below for weight measures in both the English and Metric Systems.

ENGLISH SYSTEM

Subtract: 4 lb. 9 oz.
          − 1 lb. 13 oz.

Borrow: 1 lb. = 16 oz.
Add: (16 oz. + 9 oz.
        = 25 oz.)

Write: $\overset{3}{\cancel{4}}$ lb. $\overset{2\,5}{\cancel{9}}$ oz.
        − 1 lb. 13 oz.

Answer: 2 lb. 12 oz.

Subtraction

Step 1. Since it is necessary, borrow 1 whole unit from the left column. Add this borrowed unit to the right column.
Step 2. Subtract each column.

METRIC SYSTEM

Subtract: 3 kg 750 g
           − 1 kg 900 g

Borrow: 1 kg = 1,000 g
Add: (1,000 g + 750 g
        = 1,750 g)

Write: $\overset{2}{\cancel{3}}$ kg $\overset{1,750}{\cancel{750}}$ g
        − 1 kg 900 g

Answer: 1 kg 850 g

Use borrowing to subtract the following quantities.

1.   5 ft. 3 in.   7 ft. 8 in.   5 yd. 1 ft.   2 mi.   850 yd.
    − 2 ft. 9 in.   − 3 ft. 9 in.   − 3 yd. 2 ft.   − 1 mi. 1,500 yd.

2.   8 cm 3 mm   7 cm 2 mm   13 m 52 cm   24 km 600 m
    − 5 cm 7 mm   − 3 cm 8 mm   − 9 m 85 cm   − 21 km 775 m

3.   7 lb. 8 oz.   4 lb. 5 oz.   1 lb. 12 oz.   2 T.   755 lb.
    − 3 lb. 9 oz.   − 2 lb. 8 oz.   −   14 oz.   − 1 T. 1,550 lb.

4.   5 kg 150 g   3 kg 475 g   11 g 350 mg   3 t 775 kg
    − 1 kg 875 g   − 1 kg 700 g   − 8 g 950 mg   − 1 t 825 kg

5.   6 qt.   4 qt.   4 pt. 8 oz.   2 gal. 2 qt.
    − 4 qt. 1 pt.   − 1 qt. 1 pt.   − 2 pt. 11 oz.   − 1 gal. 3 qt.

6.

2 L 350 ml	3 L 650 ml	1 L 355 ml	4 kl 325 L
− 1 L 725 ml	− 1 L 835 ml	− 850 ml	− 2 kl 750 L

7.

2 hr. 45 min.	3 hr. 25 min.	6 min. 30 sec.	3 da. 15 hr.
− 1 hr. 50 min.	− 1 hr. 45 min.	− 3 min. 35 sec.	− 1 da. 21 hr.

Solve each word problem below.

8. On Bill's third birthday, he measured 98 centimeters tall. On his sixth birthday, he stood 1 meter 23 centimeters. How much has Bill grown during these 3 years?

9. Julia is supposed to bake a beef roast for 3 hours and 15 minutes. If she has already baked it for 1 hour and 45 minutes, how much longer should she leave it in the oven?

10. When Shelley was in the sixth grade, she could throw a baseball 42 yards 2 feet. Now, in the eighth grade, she can throw it 57 yards 1 foot. How much farther can Shelley throw a baseball now than in the sixth grade?

11. While cooking, Julia started with 1 L 200 ml of milk in a pitcher. After pouring 350 ml into a bowl, how much milk did she have remaining in the pitcher?

Multiplying Measurement Units

To multiply a quantity by a number, multiply each unit of measurement separately and then simplify the answer. An example is shown below for multiplying length in both the English and Metric Systems.

ENGLISH SYSTEM

Multiply: 3 yd. 1 ft.
            ×   4
          ‾‾‾‾‾‾‾‾‾‾‾‾‾‾‾
            12 yd. 4 ft.

Answer: **13 yd. 1 ft.**
(Since 4 ft. = 1 yd. 1 ft.)

Multiplication
Step 1. Multiply each column.
Step 2. Simplify the product.

METRIC SYSTEM

Multiply: 4 cm 7 mm
            ×   3
          ‾‾‾‾‾‾‾‾‾‾‾‾‾‾‾
            12 cm 21 mm

Answer: **14 cm 1 mm**
(Since 21 mm = 2 cm 1 mm)

Multiply the following quantities and simplify each answer.

1.

5 ft. 8 in.	3 ft. 9 in.	6 yds. 2 ft.	2 mi. 880 yd.
× 4	× 3	× 4	× 2

2.

3 m 85 cm	2 m 76 cm	5 cm 6 mm	9 km 550 m
× 3	× 4	× 2	× 4

3.

2 lb. 8 oz.	3 lb. 11 oz.	3 lb. 12 oz.	1 T. 825 lb.
× 3	× 2	× 4	× 3

4.

2 kg 150 g	3 kg 400 g	5 g 225 mg	2 t 750 kg
× 8	× 6	× 5	× 2

5.

5 qt. 1 pt.	3 qt. 1 pt.	1 pt. 7 oz.	2 gal. 3 qt.
× 3	× 5	× 3	× 2

6.

2 L 350 ml	1 L 575 ml	2 L 300 ml	3 kl 650 L
× 3	× 2	× 4	× 3

7.

1 hr. 45 min.	3 hr. 25 min.	8 min. 30 sec.	1 yr. 200 da.
× 3	× 4	× 4	× 2

Solve each word problem below.

8. To make a park bench, Brian needed four iron rods, each 6 feet 8 inches long. What total length of rod did Brian need?

9. At Frank's butcher shop, Lucinda ordered eight packages of hamburger, each to weigh 1 pound 3 ounces. What total weight of hamburger will Frank use to prepare these packages?

10. In Georgia's job, she packages telephones for delivery to electronic supply stores. If each phone weighs 1 kg 250 g, what does a box that contains 12 phones weigh?

11. A recipe for ice cream calls for 1 pint 6 ounces of light cream. What total amount of light cream would be needed to make 5 times as much ice cream?

12. At his oil recycling business, Vince stores used oil in metal drums that hold 375 liters of oil. What total amount of oil can Vince store when all 12 of his storage drums are full? Express your answer in kiloliters (kl) and liters (L).

13. Renée uses pure silver to make her jewelry. If she uses strands of 4 cm 3 mm wide, how wide is a ring of 6 strands across?

Dividing Measurement Units

To divide a quantity by a number, first divide the number into the largest unit of measurement. Next, change any remainder into the next smallest unit in the problem. Then add the remainder to the smaller units already present. Finally, divide this sum by the number. An example is shown below for weight measures in both the English and Metric Systems.

ENGLISH SYSTEM

Divide:
$$3\overline{)}5 \text{ lb. } 4 \text{ oz.}$$
quotient: 1 lb.
$$\underline{3}$$
Remainder: → 2

Add: 2 lb. = 32 oz. to
second column

Answer: **1 lb. 12 oz.**
$$3\overline{)}5 \text{ lb. } 4 \text{ oz.}$$
$$\underline{3}$$
2 lb. = 32 oz.
Divide →   36 oz.
36 by 3:   36 oz.

Division

Step 1. Divide the first column.

Step 2. Change the remainder of the first column to the units of the second column. Add this remainder to the second column.

Step 3. Add the numbers in the second column. Divide this sum.

METRIC SYSTEM

Divide:
$$2\overline{)}3 \text{ kg } 250 \text{ g}$$
quotient: 1 kg
$$\underline{2}$$
Remainder: → 1

Add: 1 kg = 1,000 g
to second
column

Answer: **1 kg   625 g**
$$2\overline{)}3 \text{ kg } 250 \text{ g}$$
$$\underline{2}$$
1 kg = 1000 g
Divide →   1250 g
1250 by 2:   1250 g

Divide the following quantities.

1. $3\overline{)\,5 \text{ ft. } 3 \text{ in.}}$ $5\overline{)\,7 \text{ ft. } 6 \text{ in.}}$ $2\overline{)\,7 \text{ yd. } 1 \text{ ft.}}$ $3\overline{)\,4 \text{ mi. } 34 \text{ yd.}}$

2. $4\overline{)\,9 \text{ cm } 6 \text{ mm}}$ $3\overline{)\,7 \text{ cm } 5 \text{ mm}}$ $6\overline{)\,16 \text{ m } 26 \text{ cm}}$ $2\overline{)\,7 \text{ km } 488 \text{ m}}$

3. $2\overline{)\,5 \text{ lb. } 4 \text{ oz.}}$ $3\overline{)\,7 \text{ lb. } 5 \text{ oz.}}$ $2\overline{)\,3 \text{ T. } 600 \text{ lb.}}$ $7\overline{)\,15 \text{ kg } 470 \text{ g}}$

152

4. $3\overline{)\,5\text{ gal. 1 qt.}}$ $2\overline{)\,5\text{ gal. 2 qt.}}$ $3\overline{)\,4\text{ pt. 8 oz.}}$ $3\overline{)\,1\text{ pt. 2 oz.}}$

5. $4\overline{)\,5\text{ L 280 ml}}$ $3\overline{)\,5\text{ L 475 ml}}$ $2\overline{)\,1\text{ L 250 ml}}$ $5\overline{)\,7\text{ kl 550 L}}$

6. $2\overline{)\,3\text{ hr. 20 min.}}$ $3\overline{)\,4\text{ hr. 30 min.}}$ $2\overline{)\,5\text{ min.}}$ $2\overline{)\,3\text{ da. 8 hr.}}$

Solve each word problem below.

7. If Marie wants to cut a 6-foot 8-inch piece of ribbon into five equal pieces, how long should she cut each piece?

8. Arlene is going to cut a large piece of beef into seven rib roasts. If the uncut beef weighs 29 pounds 5 ounces, about how much should each roast weigh?

9. Measured on his bathroom scale, Julius weighs 82 kg 500 g. If Julius is three times as heavy as his son Scott, what is Scott's approximate weight in kilograms and grams?

10. Each week, Jake sets aside 9 hours during which he does his school homework. If he studies an equal amount on each day Monday through Saturday, how much does Jake study on each of these days?

Learning About Perimeter

The distance around an object is called its **perimeter**. For example, the distance around a lake is its perimeter. Perimeter is measured in length units. The symbol for perimeter is *P*.

   To determine the perimeter of a many-sided figure, add the lengths of its sides.

Example: What is the perimeter of the field at the right. Simplify the answer.

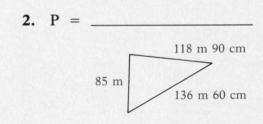

Step 1.   Add the lengths of the four sides:

12 yds. 2 ft.
9 yds. 1 ft.
14 yds. 2 ft.
+ 8 yds.
43 yds. 5 ft.

Step 2.   Simplify the answer:
a)   Change 5 ft. to 1 yd. 2 ft.
b)   Add 1 yd. 2 ft. to 43 yds.

43 yds.
+ 1 yd. 2 ft.
44 yds. 2 ft.

Answer: The perimeter is 44 yds. 2 ft.

Find the perimeter of each figure below and simplify each answer.

1. P = _____

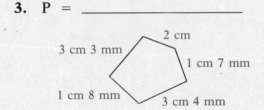

2. P = _____

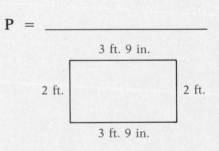

3. P = _____

4. P = _____

Word problems most often involve finding the perimeter of an object that has a common geometrical shape. The three most common geometrical shapes are the **square**, the **rectangle**, and the **triangle**.

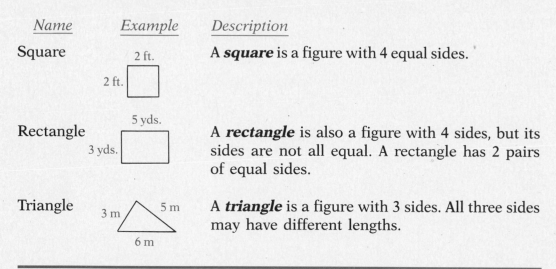

Name	Example	Description
Square	2 ft. / 2 ft.	A *square* is a figure with 4 equal sides.
Rectangle	5 yds. / 3 yds.	A *rectangle* is also a figure with 4 sides, but its sides are not all equal. A rectangle has 2 pairs of equal sides.
Triangle	3 m / 5 m / 6 m	A *triangle* is a figure with 3 sides. All three sides may have different lengths.

Using the definitions above, answer the following questions.

5. Amy's garden is in the shape of a square that measures 21 yards on each side. How many yards of fencing must Amy use to enclose the garden?

6. Allison is making a frame for a rectangular print. About how many inches of frame molding will she need if the print measures 32 inches long and 20 inches wide?

7. A piece of cut glass, in the shape of a triangle, is to be enclosed in a solder strip. What length of solder is needed if the glass sides measure 9 cm 4 mm, 12 cm 5 mm, and 14 cm?

8. Mick jogs each afternoon around Oak Park. The rectangular park is 450 yards long and 285 yards wide. How many total yards does Mick jog if he takes 5 complete laps around the park?

Becoming Familiar with Area

Area is a measure of surface. For example, to measure the size of a floor you determine its area. A larger room has more floor space and thus has a larger area than a smaller room. The symbol for area is *A*.

To measure area, you use an **area unit** in the shape of a square. A square has 4 equal sides that meet at right angles.

For example, at right is a scale drawing of a square that measures 1 foot on each side. This area unit is called a **square foot**.

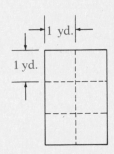

1 square foot

Common area units in the English System are the **square inch** (sq. in.), **square foot** (sq. ft.), and **square yard** (sq. yd.). Common area units in the Metric System are the **square centimeter** (cm^2) and **square meter** (m^2).

The area of a flat figure can be measured by counting the number of square area units that fit inside the figure.

Example: What is the area of the rectangle at right?

The rectangle is divided into square area units. As shown, each area unit is 1 square yard.

To find the area of this rectangle, count the number of square yards that fit inside the rectangle.

Answer: 6 square yards

Determine the area of each figure below. Be sure to include the correct area unit label as part of each answer.

1. A = _____

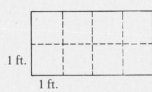

2. A = _____

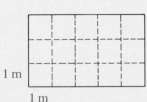

3. A = _____

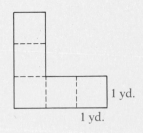

4. A = _____   **5.** A = _____   **6.** A = _____

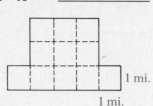

The most common area you'll ever work with is that of a rectangle. Dividing the rectangle below into 1-foot squares, you can see that the rectangle has an area of 24 square feet. Notice that we can get this answer most easily by multiplying the length (6 feet) by the width (4 feet).

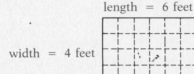

Area = 6 ft. × 4 ft. = 24 square feet

The drawing gives us a rule to use to find the area of any rectangle:

★ **To find the area of a rectangle, multiply the length by the width.**

Use the rule given above to help you answer the following questions.

7. What is the area of a room that measures 4 yards long by 3 yards wide?

8. At a cost of $12.50 per square yard, how much would it cost to carpet the room in problem #7?

9. Joan wants to tile her kitchen floor. The floor is in the shape of a rectangle 10 feet long and 7 feet wide. How many tiles will she need if each tile measures 12″ by 12″ (1 square foot)?

10. How many square yards of Astro-turf are needed to cover a football field measuring 120 yards long by 60 yards wide?

11. Guy, a groundskeeper, needs to fertilize a rectangular lawn. The lawn measures 50 meters long and 35 meters wide. If each bag of fertilizer will cover 100 square meters, how many bags will he need to complete the job?

157

Becoming Familiar with Volume

Volume is a measure of space. For example, volume is the space taken up by a solid object such as a brick. Volume is also the space enclosed by a solid surface. The volume of a box is the space enclosed by its top, its sides, and its bottom. The symbol for volume is *V*.

To measure volume, you use a **volume unit** in the shape of a cube. A cube has 6 surfaces called **faces**. Each face is a square.

At right is a scale drawing of a cube that measures 1 yard along each edge. (Each square face has 1-yard-long sides.) This volume unit is called a **cubic yard**.

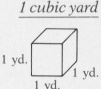

1 cubic yard

1 yd. 1 yd. 1 yd.

Common volume units in the English System are the **cubic inch** (cu. in.), **cubic foot** (cu. ft.), and **cubic yard** (cu. yd.). Common volume units in the Metric System are the **cubic centimeter** (cm^3) and **cubic meter** (m^3).

The volume of a figure can be measured by counting the number of cubic volume units that fit inside the figure.

Example: What is the volume of the rectangular solid pictured at right?

The rectangular solid is divided into cubic volume units. As shown, each volume unit is 1 cubic foot.

To find the volume of this figure, count the number of cubic feet that fit inside it.

We have numbered the 6 volume units in the front layer, and there are 3 layers.

Answer: 6 × 3 = 18 cubic feet

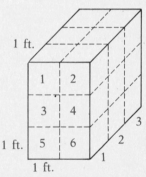

1 ft.

1 ft.   1 ft.

NOTE: There are 6 volume units in each of 3 layers.

Determine the volume of each figure below. Be sure to include the correct volume unit label as part of each answer.

1. V = _____

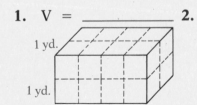

1 yd.

1 yd.

1 yd.

2. V = _____

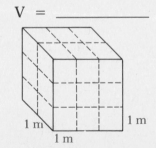

1 m   1 m

1 m

3. V = _____

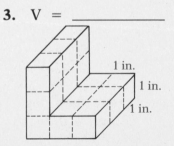

1 in.

1 in.

1 in.

The most common volume you'll ever work with is that of the rectangular solid. Boxes, suitcases, freezers, and rooms all have a rectangular solid shape. Below we have divided a rectangular solid into 24 cubic feet. Notice that we can find this volume most easily by multiplying the length (4 feet) by the width (2 feet) by the height (3 feet).

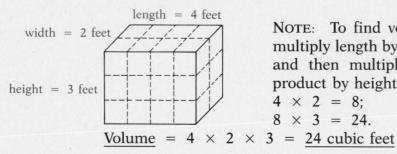

length = 4 feet
width = 2 feet
height = 3 feet

NOTE: To find volume, multiply length by width and then multiply this product by height:
4 × 2 = 8;
8 × 3 = 24.

Volume = 4 × 2 × 3 = 24 cubic feet

This drawing gives us a rule to use to find the volume of any rectangular solid:

★ **To find the volume of a rectangular solid, multiply the length by the width by the height.**

Use the rule given above to help you answer the following questions.

4. What is the volume of a storage shed that measures 4 yards long, 3 yards wide, and 3 yards high?

5. What is the volume of a small freezer that has inside dimensions of 3 feet, 2 feet, and 2 feet?

6. How many cubic feet of water are in a water bed that is 6 feet long, 5 feet wide, and 1 foot high? If 1 cubic foot of water weighs about 62 pounds, what is the weight of water in this bed when full?

7. Al's Toy Company makes plastic blocks that are in the shape of cubes 3 inches on an edge. How many of these blocks can fit in a shipping box that measures 24 inches long, 18 inches wide, and 12 inches high? (Hint: find the volume of one block and the volume of the shipping box.)

TOPIC 3:
Finding an Average

Finding an average is a good way to get a "typical" value of something. For example, suppose your phone bill for the first 3 months of the year is as follows: January, $18.25; February, $24.05; and March, $21.60. To get a value for your typical phone bill, you can compute the average of these 3 amounts.

To compute the average of a group of numbers, add the numbers together and then divide by the number of numbers in the group. Your average phone bill is easily computed:

Step 1.   Add the 3 bills.

$18.25
24.05
+ 21.60
$63.90

Step 2.   Divide by 3, the number of numbers added.

$21.30
3) $63.90

Answer: $21.30

NOTE: In most problems, as in the example, the average does not turn out to be equal to any of the numbers in the group of numbers you add. However, the average is usually close to the middle value in each group.

Answer each question below.

1. The Jenkins triplets, Bob, Ben, and Bill, weigh 124 pounds, 120 pounds, and 125 pounds respectively. What is the average weight of these three young men?

2. On a 4-day trip to Canada, Amber drove 480 miles the first day. She then drove 362 miles the second day, 412 miles the third day, and 290 miles on the fourth day. On the average, how many miles did Amber drive each day?

3. Jessica received the following scores on her four quizzes: 82, 73, 87, and 86. What is Jessica's average quiz score?

4. Three different models of a 15-inch color TV set sell for $318, $289, and $329. What is the average price of these 3 TV sets?

5. At BURGER DELIGHT, Brady sells the 5 styles of hamburgers shown at right. What is the average price of a hamburger at Brady's BURGER DELIGHT?

Delight Supreme	$3.60
Delight Delicious	2.85
Double Delight	2.25
Single Delight	1.75
Plain Delight	1.25

6. During their first 4 home basketball games, Emerald High School had the following attendance figures:

1st game: 2,112   2nd game: 1,984

3rd game: 2,080 4th game: 1,992

What is the average number of fans who attended these first 4 games?

7. Over the weekend, the Mueller family rented a VCR and 3 movies. The first movie was 1 hour 45 minutes long; the second was 2 hours 10 minutes long; and the third was 1 hour 53 minutes. What is the average length of these movies?

8. During a Valentine's Day sale, 5 local stores had the 12-ounce size of frozen orange juice concentrate on sale. Given the prices shown at right, figure out the average sale price of this concentrate.

Store	Price of Juice
Value Save	84¢
Jan's Market	79¢
Save All	81¢
Johnson's	79¢
Family Market	82¢

9. While shopping at Family Market during the Valentine's Day sale, Matt bought 3 roasts. The first weighed 6 pounds 9 ounces; the second weighed 7 pounds 11 ounces; and the third weighed 6 pounds 3 ounces. What is the average weight of these 3 roasts?

10. Part of Jed's diet plan is to keep track of the number of calories he eats each day. During the first 2 weeks, he made a list of daily calories as shown at right. Using this list, figure out the average number of daily calories Jed ate each week.

	1st Week	2nd Week
Monday	1,800	1,925
Tuesday	1,750	2,100
Wednesday	1,770	1,970
Thursday	1,810	1,975
Friday	1,800	1,900
Saturday	1,780	2,080
Sunday	1,820	1,980

1st week: _____   2nd week: _____

TOPIC 4: Squares, Cubes, and Square Roots

Learning mathematics is often compared to learning a new language. Math has many of its own words, symbols, and definitions. On these next few pages you're going to learn about **squares, cubes,** and **square roots.**

Square of a Number

The *square* of a number is that number multiplied by itself.

For example, the square of 6 is $6 \times 6 = 36$.

In symbols, we write the square of a number as a **base** and an **exponent**.

$$6 \times 6 \text{ is written } 6^2 \leftarrow \text{exponent}$$
$$\text{base}$$

The exponent (2) tells how many times the base (6) is written in the product when you multiply. The exponent of a square is always 2.

We commonly read 6^2 as "six squared." The **value** of 6^2 is 36.

Cube of a Number

The *cube* of a number is that number multiplied by itself twice.

For example, the cube of 6 is $6 \times 6 \times 6 = 36 \times 6 = 216$.

In symbols, we write a cube with an exponent of 3.

$$6 \times 6 \times 6 \text{ is written } 6^3 \leftarrow \text{exponent}$$
$$\text{base}$$

We commonly read 6^3 as "six cubed." The **value** of 6^3 is 216.

★**To find the value of the square or cube of a number, you do the multiplication represented by the symbols.**

Look at these examples:

Product	As a base and an exponent	Read in words	Value
2×2	2^2	"two squared"	4
5×5	5^2	"five squared"	25
12×12	12^2	"twelve squared"	144
$2 \times 2 \times 2$	2^3	"two cubed"	8
$4 \times 4 \times 4$	4^3	"four cubed"	64
$10 \times 10 \times 10$	10^3	"ten cubed"	1,000

Complete the following chart.

	Product	As a base and an exponent	In words
1.	4 × 4	4^2	four squared
2.	5 × 5		
3.	10 × 10		
4.	25 × 25		
5.	36 × 36		
6.	3 × 3 × 3	3^3	three cubed
7.	8 × 8 × 8		
8.	10 × 10 × 10		
9.	25 × 25 × 25		
10.	32 × 32 × 32		

Find each value as indicated below.

11. 3^2 = _____ **12.** 7^2 = _____ **13.** 9^2 = _____

14. 12^2 = _____ **15.** 15^2 = _____ **16.** 20^2 = _____

17. 3^3 = _____ **18.** 5^3 = _____ **19.** 7^3 = _____

20. 9^3 = _____ **21.** 12^3 = _____ **22.** 20^3 = _____

What Is a Square Root?

The **square root** of a number is found by asking, "What number times itself equals this number?" For example, to find the square root of 25 we ask, "What number times itself equals 25?"
The answer is 5 because $5 \times 5 = 25$.

The symbol for square root is $\sqrt{}$. Thus, $5 = \sqrt{25}$.

Numbers that have whole number square roots are called **perfect squares**. A list of perfect squares is easily made by "squaring" whole numbers. The first 15 perfect squares are shown in the table below.

Table of Perfect Squares

$1^2 = 1$	$6^2 = 36$	$11^2 = 121$
$2^2 = 4$	$7^2 = 49$	$12^2 = 144$
$3^2 = 9$	$8^2 = 64$	$13^2 = 169$
$4^2 = 16$	$9^2 = 81$	$14^2 = 196$
$5^2 = 25$	$10^2 = 100$	$15^2 = 225$

The values in this table can be used to find the square roots of perfect squares.

Example: What is the square root of 169?

Find 169 in the column at right.

Since $13^2 = 169$, 13 is the square root of 169.

In symbols, $13 = \sqrt{169}$.

Write each sentence below in symbols. The first one is done as an example.

1. Four is the square root of sixteen. $4 = \sqrt{16}$

2. Ten is the square root of one hundred. _____

3. Nine is the square root of eighty-one. _____

4. Twelve is the square root of one hundred forty-four. _____

Use the Table of Perfect Squares to find each square root below.

5. $\sqrt{64}$ = ____

6. $\sqrt{4}$ = ____

7. $\sqrt{121}$ = ____

8. $\sqrt{25}$ = ____

9. $\sqrt{49}$ = ____

10. $\sqrt{225}$ = ____

11. $\sqrt{1}$ = ____

12. $\sqrt{196}$ = ____

13. $\sqrt{36}$ = ____

Book 1 Post-Test

The following 60 questions will give you a chance to review briefly many of the skills you've learned in Book 1. The questions are divided into the four computation skills and word-problem skills.

Work carefully and answer every question. When you finish, check your answers with the answers given on pages 185–186.

ADDITION SKILLS

1.   5
     +3

2.   7
     +0

3.   4
     2
     +3

4. 6 + 2 + 0 =

5.   26
     +13

6.   31
     23
     +12

7.   58
     +29

8.   75
     +49

9. $6.37
     + 2.74

10.   675
     359
     + 63

11.   4,048
     2,349
     + 378

12. 4,274 + 3,138 =

SUBTRACTION SKILLS

13.   8
     −5

14.   17
     − 9

15.   68
     −35

16.   35
     −18

17.   875
     − 89

18. $23.48
     − 10.09

19.   670
     − 58

20.   408
     −256

21.   900
     −549

22. 800 − 378 =

23. $500.00 − $453.63 =

MULTIPLICATION SKILLS

24.   8
     ×5

25.   53
     × 2

26.   23
     ×12

27.   321
     ×123

28.   52
     ×30

29.   530
     ×201

30.   47
     × 8

31.   98
     ×73

32. 678
 $\times\,607$

33. $236 \times 19 =$

34. $4{,}508 \times 1{,}000 =$

DIVISION SKILLS

35. $9\overline{)45}$

36. $3\overline{)96}$

37. $3\overline{)6{,}369}$

38. $2\overline{)60}$

39. $3\overline{)627}$

40. $5\overline{)4{,}515}$

41. $6\overline{)39}$

42. $7\overline{)94}$

43. $4\overline{)\$1.32}$

44. $12\overline{)360}$

45. $18\overline{)414}$

46. $25\overline{)5{,}325}$

47. $279\overline{)2{,}232}$

48. $240 \div 16 =$

49. $7{,}973 \div 12 =$

WORD-PROBLEM SKILLS

Solve each of the following word problems. For each problem, circle the letter of the correct answer from the answer choices given.

50. Bobby had 4 coins in his pocket: a fifty-cent piece, a quarter, a dime, and a penny. How much money did Bobby have in change?

a) $.61
b) $.81
c) $.86
d) $.91
e) $.98

51. Traveling by plane, Dallas is 795 miles from Atlanta. Also by plane, Dallas is 936 miles from Minneapolis. How many miles farther is Dallas from Minneapolis than from Atlanta?

a) 141
b) 143
c) 259
d) 459
e) 1,731

52. On Monday's shopping trip, Marilyn bought a lamp for $23.49, a table for $49.50, 2 vases for $19.95 apiece, and a serving plate for $12.79. What total amount did Marilyn pay for the lamp and the 2 vases?

a) $ 43.44
b) $ 63.39
c) $ 66.93
d) $105.73
e) $125.68

53. At a cost of $1.49 per bottle, how much do 21 bottles of cola cost?

a) $14.90
b) $21.29
c) $30.00
d) $31.29
e) $41.29

54. While dieting, Ruth lost 4 pounds during May and 5 pounds during June. How many pounds did she lose altogether if her weight dropped from 145 pounds on May 1 to 131 pounds on August 1?

a) 5
b) 9
c) 14
d) 18
e) 23

55. Allison applied for a typing job that required a typing speed of at least 65 words per minute. She was given a 6-minute timed typing test. On this test, she typed 348 words. How many words per minute did Allison average on this test?

a) 58
b) 59
c) 65
d) 390
e) 2088

56. At a hardware sale, Roy bought a drill on sale for $39.95 and six screwdrivers on sale for 99¢ each. How much change should Roy receive if he pays for these items with two twenty-dollar bills and a ten-dollar bill?

a) $ 1.04
b) $ 4.11
c) $ 5.94
d) $ 9.06
e) $14.11

57. A recipe for tangy punch calls for 2 qt. 1 pt. of lime soda. This will be enough punch for 8 large servings. If Jill wants to make 6 times this much punch, how much lime soda should she use? Express your answer in the largest units possible.

a) 3 gal. 6 pt.
b) 3 gal. 3 qt.
c) 4 gal. 1 qt.
d) 6 gal. 2 qt.
e) 12 qt. 6 pt.

58. A roll of R-30 insulation covers 64 square feet of surface. If each roll costs $25.50, how much will it cost to insulate a ceiling that is in the shape of a rectangle measuring 40 feet long by 32 feet wide?

a) $ 51.00
b) $125.00
c) $255.00
d) $375.00
e) $510.00

59. During league play on Saturday night, Ellen bowled 4 games, all above her year-long average of 142. The scores of her 4 games were 144, 151, 167, and 146. What was Ellen's average score for just these 4 games?

a) 142
b) 144
c) 145
d) 148
e) 152

60. Wilma has a large storage freezer in the shape of a rectangular solid. The inside dimensions of this freezer are as follows: length = 4 feet; width = 2 feet; and height = 3 feet. How many cubic feet of storage space does this freezer contain?

a) 9
b) 24
c) 81
d) 124
e) 234

Book I Post-Test Evaluation Chart

On the chart below, circle the number of any problem you missed. The skill and study pages associated with each problem are indicated.

Problem Numbers	Associated Skills	Study Pages
ADDITION SKILLS		
1, 2, 3, 4, 5, 6	Adding small numbers (no carrying)	18 to 23
7, 8, 9, 10, 11, 12	Adding and carrying	32 to 36
SUBTRACTION SKILLS		
13, 14, 15	Subtracting small numbers (no borrowing)	44 to 47
16, 17, 18	Subtracting and borrowing	56 to 59
19, 20, 21, 22, 23	Borrowing from zeros	60 to 63
MULTIPLICATION SKILLS		
24, 25, 26, 27, 28, 29	Multiplying (no carrying)	72 to 83
30, 31, 32, 33, 34	Multiplying and carrying	84 to 90
DIVISION SKILLS		
35, 36, 37, 38	Short division	100 to 104
39, 40	Using zero as a place holder	107
41, 42, 43, 44, 45, 46, 47, 48, 49	Long division	110 to 111

WORD-PROBLEM SKILLS

In addition to the page references given here, there are other pages in which word-problem skills are emphasized. If you miss any of the problems below, make sure you work carefully through all of the pages on word problems.

50, 52	Addition word problems	30
51, 54	Subtraction word problems	51
53	Multiplication word problem	82
55	Division word problem	125
56	Multi-step word problem	134 to 135
57	Measurement units word problem	140 to 159
58	Measurement: area word problem	
60	Measurement: volume word problem	
59	Averages word problem	160 to 161

Answer Key

Overview of Skills
pages 4–5

1. 7	9. 5	17. 28	25. 6
2. 9	10. 32	18. 96	26. 731
3. 7	11. 67	19. 528	27. 20
4. 64	12. 414	20. 860	28. 304
5. 39	13. 451	21. 220,460	29. 25
6. 64	14. 133	22. 468	30. 32 r 4
7. 1,267	15. 631	23. 3,724	31. 209
8. 6,215	16. 587	24. 74,305	32. 23

33.
```
   174
   250
 + 195
   619
```

34.
```
  $20.00
 −$13.39
 $ 6.61
```

35.
```
   $1.03
 ×    12
    206
    103
  $12.36
```

36.
```
         14 months
 $24) $336
       − 24
         96
       − 96
          0
```

37.
```
  $61.00
 +$34.64
  $95.64
```

38.
```
  1 lb.  7 oz.
 −      11 oz.
        12 oz.
```

39. STEP 1:
```
     45 min.
   × 7 days
   315 min.
```
STEP 2:
```
              5 hrs. 15 min.
  60 min.) 315 min.
           − 300
              15
```

40.
```
    7 yards
  × 4 yards
  28 square yards
```

41. STEP 1: 6 × 6 × 10 = 360 cu. feet
STEP 2:
```
                360 boxes
    1 cu. ft.) 360 cu. ft.
```

42. STEP 1:
```
      98
     111
   + 106
     315
```

STEP 2:
```
      105 average score
  3) 315
    − 3
     015
    − 15
       0
```

Chapter 1:   Whole Numbers

Digits and Place Value
page 8

1. eighty
2. thirty
3. five
4. three thousand
5. eight hundred
6. twenty thousand
7. two hundred thousand
8. zero or no hundreds
9. three million

Reading Whole Numbers
page 9

1. 1, 8	6. 7, 8, 2	11. 8, 0, 9, 3
2. 2, 5	7. 3, 0, 9	12. 3, 4, 0, 6
3. 3, 6	8. 7, 2, 0	13. 5, 4, 2
4. 4, 0	9. 4, 9, 8, 2	14. 8, 0, 9
5. 1, 3, 7	10. 5, 7, 3, 1	15. 4,7,0

Writing Whole Numbers
page 10

1. forty-seven
2. thirty-nine
3. fifty-six
4. ninety-two
5. two hundred thirty-eight
6. five hundred seventy-nine
7. seven hundred thirty-five
8. eight hundred thirty
9. three thousand, five hundred eighty-seven
10. five thousand, eight hundred sixty-five
11. seven thousand, three hundred twenty-nine
12. twenty-three thousand, two hundred fifty-four
13. thirty-eight thousand, six hundred seventy-five
14. seventy-three thousand, one hundred eighty-four

Writing Zero as a Place Holder
page 11

1. 308	6. 604,200
2. 507	7. 817,605
3. 3,015	8. 6,307,000
4. 46,011	9. 8,201,000
5. 209,400	10. 9,702,013

Rounding Whole Numbers
pages 12–13

1. 50, 90, 80, 120
2. 200, 500, 2,500

3. 5,000, 6,000, 13,000
4. 30, 80, 130, $280
5. 200, $400, 4,700, $5,400
6. 5,000, 5,000, $9,000, $12,000
7. $1,300
8. 70
9. 93,000,000
10. $79,000

Working with Dollars and Cents
pages 14–15

1. $20.08	10. $87.55
2. $32.09	11. 8¢, $0.08
3. $61.15	12. 9¢, $0.09
4. $18.38	13. 4¢, $0.04
5. $52.03	14. 3¢, $0.03
6. $25.75	15. 35¢, $0.35
7. $115.50	16. 43¢, $0.43
8. $79.80	17. 98¢, $0.98
9. $309.04	18. 57¢, $0.57

19. five dollars and seven cents
20. twelve dollars and six cents
21. seven dollars and thirty-seven cents
22. twenty-three dollars and sixty-five cents
23. one hundred twenty-five dollars and fifty cents

Rounding Dollars and Cents
page 16
1. $5.80, $1.60, $8.10, $10.30
2. $7.00, $13.00, $53.00, $9.00
3. $0.90
4. $0.70

Reading, Writing, and Rounding Skills Review
page 17
1. ninety, two thousand, six, five hundred
2. 5, 0, 3, 8
3. 0, 9, 2, 7
4. one hundred sixty-three
5. four thousand, seventy-nine
6. twenty-seven thousand

7. 575	12. 3,000
8. 2,046	13. $0.40
9. 4,600,000	14. $3.00
10. 80	15. $70.00
11. 200	

Chapter 2: Addition
Concepts in Addition
pages 18–19
1. 7, 7, 9, 9, $16, $4.93
2. yes
3. 12 boats + 9 boats
4. $5.60 + $3.45
5. 50 + 19
6. 13 + 7
7. 9 bottles + 17 bottles
8. $4.50 + $3.16

Adding Single Digits
pages 20–21
1. 8, 7, 3, 2, 6, 6, 4
2. 1, 9, 7, 7, 8, 9, 4
3. 7, 9, 6, 5, 6, 9, 0
4. 11, 10, 14, 13, 13, 10, 13
5. 15, 11, 15, 15, 11, 13, 10
6. 12, 14, 12, 15, 13, 13, 11
7. 12, 12, 14, 10, 13, 16, 10

8. A.		B.	
1. f		1. e	
2. c		2. h	
3. e		3. a	
4. h		4. f	
5. g		5. g	
6. b		6. d	
7. d		7. b	
8. a		8. c	

9.
```
  9 boards       7        3¢
+ 4 boards      +8       +9¢
 13 boards      15       12¢
```
10.
```
   0       8 nails      $4
  +6      +5 nails     + 2
   6      13 nails      $6
```
11.
```
   5       3       8¢
  +7      +0      +6¢
  12       3      14¢
```
12.
```
   5       8       7 buttons
  +9      +4      +1 button
  14      12       8 buttons
```

Adding a Column of Digits
page 22
1. 8, 9, 8, 7, 18, 15, 15
2. 17, 12, 24, 17, 12, 15, 21
3. 13, 20, 13, 25, 22, 21, 15
4. 16, 16, 15, 19, 18, 17, 17
5. 16, 19, 20, 16, 20, 17, 16

Adding a Row of Digits
page 23

1. 15, 17, 9	5. 12, 17, 17
2. 16, 17, 15	6. 17, 21
3. 9, 14, 19	7. 16, 21
4. 13, 14, 17	8. 16, 15

9. 6 + 7 + 9 = 22
10. $8 + $9 + $2 = $19
11. 8 gallons + 4 gallons + 7 gallons = 19 gallons
12. 7 + 6 + 9 + 8 = 30

Adding Larger Numbers
page 24
1. 38, 88, 84, 96, 59, 29
2. 65, 98, 85, 39, 98, 67
3. 789, 719, 972, 439, 978, 589
4. 85, 66, 59, 39, 48, 89

Adding with Empty Columns
page 25
1. 19, 46, 28, 287, 397
2. 498, 779, 496, 487, 486
3. 3,756, 5,798, 5,899, 7,788, 5,658
4. 8,975, 27,687, 38,878, 3,556, 29,957

5. 49, 19, 259, 236, 775
6. 2,748, 4,762, 9,679, 14,979, 28,687

Adding Dollars and Cents
pages 26–27
1. $7.58, $4.58, $8.75, $7.85, $2.59, $9.79
2. $14.83, $44.59, $55.97, $29.58, $35.97, $35.25
3. 39¢ or $0.39, 79¢ or $0.79, 98¢ or $0.98
4. $4.95, $1.56, $1.85, $2.89, $3.95
5. 76¢ $0.76, 70¢ $0.70, 18¢ $0.18
6. 37¢ $0.37, 78¢ $0.78, 65¢ $0.65
7. $32.35, $15.56, $26.17
8. $8.61, $21.85, $36.45

Introducing Word Problems
pages 28–29
1. **Question:** How much rent does she pay each month?
   Find: monthly rent
2. **Question:** How much do two gallons of milk weigh?
   Find: weight of 2 gallons
3. **Question:** What amount did Manuel pay in state and federal taxes?
   Find: federal and state taxes
4. **Question:** How many items did Joan buy?
   Find: number of items
5. **Question:** Driving through Rockville, how many miles is it between Oak Grove and Salem?
   Find: miles from Oak Grove to Salem
6. **Find:** number of cats and dogs
7. **Find:** cost of skirt and blouse
8. **Find:** total overtime pay

Addition Word Problems
pages 30–31
1. **Find:** How many students are in the class?
   Solution: 23 men
   <u>16 women</u>
   39 students
2. **Find:** How far is Seattle from Eugene?
   Solution: 172 miles
   <u>+114 miles</u>
   286 miles
3. **Find:** cost of 3 candy bars
   Solution: 23¢
   23¢
   <u>+23¢</u>
   69¢
4. **Find:** How many screwdrivers does George have altogether?
   Solution: 4 screwdrivers
   <u>+5 screwdrivers</u>
   9 screwdrivers
5. **Find:** the purchase price of car
   Solution: $6,250
   <u>+ 1,245</u>
   $7,495

6. **Find:** How much money does Tina have in change?
   Solution: 25¢
   10¢
   10¢
   10¢
   01¢
   01¢
   01¢
   <u>+01¢</u>
   59¢
7. **Find:** What time will it be in 27 minutes?
   Solution: 12 minutes
   <u>+27 minutes</u>
   39 minutes;   8:39
8. **Find:** What take-home monthly income does he need to pay these expenses?
   Solution: $415
   <u>+ 581</u>
   $996
9. **Find:** What was the temperature at noon?
   Solution: 43°F
   <u>+14°F</u>
   57°F
10. **Find:** How much did Mrs. Murphy pay for 2 hats?
    Solution: $8.25
    <u>+ 9.50</u>
    $17.75
11. **Find:** How much did Barbara weigh on her fifteenth birthday?
    Solution: 101 pounds
    <u>+ 14 pounds</u>
    115 pounds
12. **Find:** cost of a 3-minute call
    Solution: 30¢
    21¢
    <u>+21¢</u>
    72¢
13. **Find:** How many inches of rain fell on the first three days of the week?
    Solution: 3 inches
    5 inches
    <u>+4 inches</u>
    12 inches
14. **Find:** total of John's bills
    Solution: $405.00
    21.14
    <u>+ 71.25</u>
    $497.39
15. **Find:** combined attendance for these first 3 games
    Solution: 2,324 people
    3,042 people
    <u>+3,121 people</u>
    8,487 people

Adding and Carrying
pages 32–35
1. 33, 54, 92, 131, 76, 82
2. 55, 72, 60, 105, 125, 64
3. $42, 35¢, $51, $93, 81¢, 70¢
4. 22, 47, 49, 55, 83, 132
5. 409, 635, 917, 1,228, 429, 1,434
6. 336, 978, 528, 509, 1,518, 1,127
7. 759, 467, 417, 1,145, 1,219, 2,426
8. 9,183, 5,297, 8,484, 11,664, 9,278, 11,235
9. 5,486, 6,388, 4,579, 9,297, 13,266, 16,294
10. 8,268, 4,078, 5,376, 11,369, 10,468, 23,378
11. 621, 422, 930, 1,315, 953, 1,427
12. 761, 504, 202, 662, 312, 526
13. 841, 604, 870, 950, 920, 971
14. 1,333, 1,502, 1,273, 1,200, 1,507, 1,110
15. 455, 585, 422, 1,820, 1,625, 2,515
16. 4,338, 9,601, 12,638, 8,925, 9,167
17. 5,229, 6,519, 9,120, 4,162, 6,261
18. 8,329, 9,231, 9,251, 10,359, 11,830
19. 3,087, 7,342, 7,237, 13,214, 16,150
20. 21,225, 22,311, 73,200, 135,707, 105,180
21. 41,069, 63,154, 34,451, 57,281, 72,440
22. 20,939, 28,685, 73,200, 101,390, 144,576

Adding Numbers in a Row
page 36
1. 37, 23, 28
2. 68, 83, 77
3. 233, 258, 629
4. 1,435, 1,551, 4,069
5. 9,703, 36,709

Carrying with Dollars and Cents
page 37
1. $10.52, $13.66, $18.11, $43.21, $40.60, $22.23
2. $3.32, $4.72, $11.49, $15.54
3. $118
4. $39.28

Finding Necessary Information
pages 38–39
1. **Question:** What is the total price of a quart of milk and a loaf of bread?
   Necessary information: $0.48, $1.09
2. **Question:** What amount did she spend for clothes?
   Necessary information: $34.95, $19
3. **Question:** How much does Brenda want to lose in all?
   Necessary information: 13 pounds, 14 pounds
4. **Question:** How many total hours does she work each week?
   Necessary information: 35 hours, 8 hours
5. **Question:** What was the total rainfall during the first 3 days of the week?
   Necessary information: 2 inches, 1 inch, 2 inches
6. **Question:** What is the sum of his 3 Tuesday night scores?
   Necessary information: 147, 172, 165
7. **Question:** How many people voted in this election?
   Necessary information: 4,572, 6,385
8. **Question:** What is the total cost of a 7-lb. package of chicken?
   Necessary information: 7 pounds, $.89/lb.

Addition Word Problems
pages 40–41
1. 12 miles
    9 miles
    + 13 miles
    34 miles
2. 9 pounds
    + 11 pounds
    20 pounds
3. $ 6.95
    + 14.95
    $21.90
4. $36.00
    + 24.75
    $60.75
5. 6,284 people
    5,782 people
    + 800 people
    12,866 people
6. 209 miles
    248 miles
    + 173 miles
    630 miles
7. $21.00
    + 8.75
    $29.75
8. 30¢
    13¢
    + 21¢
    64¢
9. 3 pounds
    4 pounds
    + 5 pounds
    12 pounds
10. $ 9.49
    + 13.50
    $22.99
11. 14 years
    + 6 years
    20 years old
12. $75.50
    75.50
    75.50
    $226.50
13. $2.49
    .65
    .75
    + .21
    $4.10
14. $1,475
    276
    + 125
    $1,876

Addition Skills Review
pages 42–43
1. 8, 15, 19, 12, 19
2. 39, 99, 195, 778, 678, 4,797
3. 33, 52, 263, 543, 846, 3,018
4. 131, 144, 285, 1,104, 2,574, 27,389
5. $5.33, $6.85, $9.85, $171.68, $4,115.07
6. 24, 56, 63, 1,149
7. 55, 101, 532
8. 23
9. $31.00
10. 105 pounds
11. $0.82
12. $3.30
13. $1.43
14. $.50
    .30
    + .07
    $.87
15. 9
    7
    + 3
    19

16. $9.89   17.   32,640   18.   3,476
        4.95   17,456   +2,890
    + 2.39   +16,540   6,366
   $17.23   66,636

Chapter 3: Subtraction

Concepts in Subtraction
pages 44–45

1. 9, 7, 13, 25, $12, $27.35
2. no
3. 30 − 12
4. 19 − 6
5. 142 − 77
6. 87 pounds − 60 pounds
7. 59 − 33
8. 97 feet − 40 feet

Basic Subtraction Facts
pages 46–47

1. 2, 3, 2, 4, 1, 4, 4, 0
2. 7, 4, 3, 2, 8, 1, 6, 3
3. 1, 0, 5, 3, 1, 0, 1, 2
4. 6, 1, 3, 0, 9, 2, 5, 2
5. 5, 7, 5, 9, 4, 3, 7, 6
6. 2, 5, 6, 7, 9, 9, 9, 5
7. 5, 3, 6, 3, 7, 8, 4, 6
8. 2, 4, 7, 8, 7, 4, 9, 8
9. 9, 9, 9, 8, 8, 6, 7, 6
10. A. 1. f   B. 1. h
       2. h   2. d
       3. g   3. f
       4. a   4. b
       5. c   5. g
       6. b   6. a
       7. d   7. c
       8. e   8. e
11.   6 buttons   12 $.10
     −3 buttons − 4 − .05
     3 buttons   **8**   **$.05**
12.   7   14 videotapes   12
    −3   − 6 videotapes − 7
    4   **8 videotapes**   **5**
13. $11.00   15 cars   9
    − 7.00 − 9 cars −7
    $ 4.00   **6 cars**   **2**
14.   18 tires   11 $13.00
    − 9 tires − 4 − 8.00
    9 tires   **7**   **$ 5.00**

Subtracting Larger Numbers
page 48

1. 12, 11, 34, 21, 15, 11, 33
2. 32, 12, 22, 13, 22, 41, 17
3. 123, 314, 212, 221, 611, 111, 43
4. 13, 22, 11, 32, 20, 41, 13
5. 141, 212, 425, 134, 311, 522, 612
6. 2,423, 7,815, 3,813, 3,232, 2,251, 3,424, 1,215

Writing Zeros in the Answer
page 49

1. 20, 20, 6, 10, 6, 30, 30

2. 302, 606, 102, 410, 201, 41, 610
3. 4,013, 1,200, 2,103, 5,062, 123, 3,035, 2,400
4. 1,103, 2,103, 580, 11,003, 11,052, 10,124, 1,215
5.   170   7.   95¢
    − 60   −65¢
    110   **30¢**
6.   598 miles   8.   250 pounds
   −296 miles   −220 pounds
   302 miles   **30 pounds**

Subtracting Dollars and Cents
page 50

1. $2.64, $3.34, $1.23, $ .25, $2.54, $ .26
2. $14.35, $31.20, $12.25, $3.21, $13.12, $4.21
3. $14.88   6. $1.98
    − 2.57   − .47
    $12.31   **$1.51**
4. $3.99   7. $275.85
    − .75   − 124.50
    $3.24   **$151.35**
5. $9.99   8. $0.98
    − 5.00   − .37
    $4.99   **$ .61**

Subtraction Word Problems
page 51

1.   1986 years   5.   197 pounds
   −1941 years   −161 pounds
   45 years   **36 pounds**
2. $39.99   6. $895
    − 25.50   − 375
    $14.49   **$520**
3.   19 inches   7.   228 miles
   − 5 inches   −115 miles
   14 inches   **113 miles**
4.   67°   8.   3:46 P.M.
   −13°   − 30
   54°   **3:16 P.M.**

Recognizing Key Words
pages 52–53

1. **Key word:** change; subtraction
2. **Key words:** in all; addition
3. **Key word:** farther; subtraction
4. **Key words:** difference, between; subtraction
5. **Key words:** together, combined; addition
6. **Key words:** more than; subtraction
7. **Key word:** reduce; subtraction
8. **Key words:** increased, total; addition
9. **Key word:** more; addition
10. **Key words:** older than; subtraction

Solving Word Problems
pages 54–55

1. $395
    − 320
    $ 75

2. 75¢
    −64¢
    11¢

3. $575.43
    + 325.00
    $900.43

4. $17.00
    47.25
    + 1.75
    $66.00

5. 57,863 miles
    − 50,000 miles
    7,863 miles

6. $435.00
    125.00
    147.34
    + 67.87
    $775.21

7. $127.85
    − 106.00
    $ 21.85

8. 71°
    +22°
    93°

9. $3.29
    − 2.17
    $1.12

10. 25 hours
    + 6 hours
    31 hours

11. $2,945
    − 825
    $2,120

12. 8 pounds
    7 pounds
    +5 pounds
    20 pounds

13. 3,955 pounds
    −2,630 pounds
    1,325 pounds

14. $25.00
    45.50
    + 27.75
    $98.25

15. 1,771 miles
    − 802 miles
    969 miles

Subtracting by Borrowing
pages 56–57
1. 15, 28, 28, 17, 109, 116, 236
2. 38, 14, 57, 27, 8, 17, 34
3. 17, 139, 219, 717, 365, 347, 508
4. 243, 755, 488, 373, 299, 72, 186
5. 56, 274, 342, 250, 871, 771, 284
6. 81, 283, 83, 281, 54, 123, 175
7. 3,732, 1,613, 435, 1,811, 3,934, 733
8. 4,811, 3,863, 2,850, 6,541, 723, 1,912
9. 1,622, 2,535, 624, 1,762, 10,422, 1,539

Borrowing from Two or More Columns
pages 58–59
1. 387, 568, 678, 285, 589, 269, 25
2. 558, 766, 599, 282, 288, 68, 759
3. 287, 386, 175, 389, 53, 587, 188
4. 18, 388, 2,823, 593, 2,865, 6,586
5. 17, 23, 36, 8, 172, 329, 68
6. 83, 247, 488, 6,722, 1,930, 3,781
7. 2,679, 4,189, 2,578, 648, 3,787, 3,888
8. 1,886, 905, 1,492, 2,588, 5,137, 14,518

Subtracting from Zeros
page 60
1. 356, 462, 39, 254, 3,027, 4,214
2. 459, 37, 132, 158, 383, 221
3. 76, 567, 7,211, 2,658, 4,069, 677

Subtracting from Separated Zeros
page 61
1. 4,287, 3,185, 2,184, 721
2. 3,452, 6,305, 2,817, 7,396, 2,307

Subtracting from a Row of Zeros
page 62
1. 211, 247, 414, 4,513, 1,236, 12,145
2. 155, 427, 774, 124, 408, 247
3. 2,160, 1,050, 1,595, 2,225, 15,260, 6,031

Subtracting Numbers in a Row
page 63
1. 61, 252, 1,336
2. 118, 34, 1,524
3. 275, 2,688, 149
4. 633, 111, 227
5. 553, 571, 4,264

Borrowing with Dollars and Cents
page 64
1. $2.17, $3.87, $3.25, $26.87, $78.67, $109.55
2. $2.63, $3.42, $7.63, $28.17, $157.35, $110.01
3. $4.99, $ .71, $1.72, $4.37
4. $14.65
5. $24.85

Deciding When to Add or Subtract
pages 65–66
1. b   4. a   7. a
2. b   5. b   8. a
3. b   6. b   9. b

Solving Word Problems
pages 67–69

1. $1.00
    − .79
    $ .21

2. 3,145 pounds
    − 876 pounds
    2,269 pounds

3. $837.87
    + 125.75
    $963.62

4. 3:00 P.M.
    +2 hours
    5:00 P.M.

5. $415
    − 350
    $ 65

6. 73 degrees
    −68 degrees
    5 degrees

7. 1915
    + 80
    1995

8. 1,760 yards
    −1,100 yards
    660 yards

9. 954 miles
    876 miles
    +1,105 miles
    2,935 miles

10. 2,350 calories
    − 500 calories
    1,850 calories

11. 280 calories
    −120 calories
    160 calories

12. $7.95
    2.98
    2.98
    1.20
    1.20
    + .48
    $16.79

13. $9.99
    − 2.00
    $7.99

14. $960
    − 720
    $240

15. $275
    137
    + 156
    $568

16. $275
    − 156
    $119

17. 3,225 calories
    − 750 calories
    2,475 calories

18. 214 pounds
　 − 196 pounds
　　 18 pounds

19. $2,845
　 +　 735
　　 $3,580

Subtraction Skills Review
pages 70–71

1. 3, 0, 11, 6, 3, 7
2. 23, 16, 21, 620, 301, 2,233
3. 27, 28, 129, 207, 184, 664
4. 292, 666, 2,219, 2,611, 3,483, 6,417
5. $4.10, $1.13, $5.07, $5.89, $2.17, $11.51
6. 9, 28, 15, 119
7. 86, 24, 263, 2,626
8. 7
9. $38.00
10. 29 inches
11. $.55
12. $3.29
13. $.63
14. 　 $1.00
　　 −　 .89
　　　 $ .11
15. 　 $2.00
　　 −　 .84
　　　 $1.16
16. 　 738 miles
　　 − 668 miles
　　　 70 miles
17. 　 $289
　　 − 239
　　　 $ 50
18. 　 5:00 P.M.
　　 − 3 　 hours
　　　 2:00 P.M.

Chapter 4: Multiplication
Concepts in Multiplication
pages 72–73

1. 15, $.09, 7, 8, 0, 0
2. 3 × 6 and 6 + 6 + 6
3. 249 lbs. × 18
4. 40 × 23
5. $5.18 × 6
6. 400 miles × 7

Basic Multiplication Facts
pages 74–75

1. 18, 56, 9, 0, 12, 27, 56, 5
2. 0, 6, 35, 24, 30, 0, 72, 54
3. 28, 48, 32, 56, 21, 36, 40, 10
4. 4, 20, 32, 21, 63, 72, 24, 15
5. 12, 40, 16, 18, 42, 16, 45, 18
6. 30, 27, 9, 10, 14, 48, 12, 16
7. 27, 6, 49, 42, 45, 64, 15, 35
8. 18, 24, 12, 63, 6, 28, 36, 36
9. 40, 9, 20, 56, 6, 18, 24, 2
10. $8 × 5 = $40, 7 × 6 = 42, 9 × 8 = 72
11. 3 lbs. × 9 = 27 lbs.,
　 5 yards × 6 = 30 yds., 4 × 7 = 28

Multiplying by One Digit
page 76

1. 63, 64, 88, 68, 39, 80, 36
2. 328, 188, 128, 426, 720, 249, 567
3. 936, 282, 696, 806, 639, 648, 480
4. 1,293, 2,169, 4,824, 2,884, 1,296, 7,288, 1,890

Multiplying by Two Digits
page 77

1. 682, 690, 480, 1,196, 5,880, 12,504, 18,146
2. 384, 528, 341, 840, 1,517, 18,103, 16,384

Multiplying by Three Digits
page 78

1. 15,972, 68,586, 40,610, 88,842, 374,400, 240,751, 494,373
2. 36,288, 25,531, 67,731, 89,676, 13,888, 48,760, 36,300
3. 158,600, 530,400, 964,100, 139,103, 678,824, 970,944, 869,320

Zeros in Multiplication
page 79

1. 460, 430, 420, 2,040, 3,960, 28,480, 12,390
2. 67,410, 42,240, 44,520, 14,300, 30,800, 44,730, 37,440
3. 48,600, 168,400, 183,600, 100,400, 21,836, 83,839, 126,721

Multiplying by 10, 100, and 1,000
page 80

1. 250, 670, 590, 1,350, 3,520, 5,890
2. 12,700, 25,400, 63,400, 35,200, 48,500, 38,600
3. 8,500, 90, 7,600, 70, 36,500, 140
4. 36,000, 65,000, 375,000, 634,000, 578,000, 4,586,000

Multiplying Dollars and Cents
page 81

1. $15.90, $16.80, $27.96, $168.88, $12.48, $28.80
2. $585.00, $736.00, $75.80, $650.00, $326.90, $7,490.00
3. 　 $ .92
　　 ×　 4
　　 $3.68
　　 　 100
　　 × .74
　　 $74.00
　　 　 $ .53
　　 ×　 3
　　 $1.59

Multiplication Word Problems
pages 82–83

1. **Key word:** total
　 　 $4.00
　 ×　 21
　 　 4.00
　 　 80.0
　 $84.00
2. 　 52 miles
　 ×　 4
　 208 miles
3. **Key words:** in all
　 　 $42
　 ×　 12
　 　 84
　 　 42
　 $504
4. **Key word:** twice
　 　 $ .63
　 ×　 2
　 $1.26
5. 　 13
　 ×　 2
　 26 yards
6. 　 12
　 ×　 4
　 48 pounds
7. **Key word:** multiply
　 　 11
　 ×　 8
　 88 square feet
8. 　 12
　 ×　 4
　 48 hours

Multiplying and Carrying
pages 84–87

1. 84, 76, 518, 416, 378, 1,080, 2,187
2. 34, 75, 96, 108, 94, 87, 76
3. 378, 222, 210, 600, 432, 156, 111
4. 274, 384, 430, 387, 2,468, 1,284, 1,438
5. 624, 728, 981, 3;248, 3,228, 1,830

6. 535, 824, 921, 4,942, 7,227, 3,654
7. 324, 843, 810, 1,746, 3,927, 5,920
8. 813, 608, 1,110, 768, 753, 568
9. 1,928, 1,142, 5,580, 3,227, 3,480, 1,405
10. 19,860, 7,269, 8,106, 6,448, 17,169, 27,284
11. 5,648, 7,863, 3,626, 8,730, 7,222, 5,139
12. 22,884, 13,896, 65,680, 23,169, 67,270, 15,848
13. 725, 717, 712, 1,068, 5,265, 5,929
14. 1,002, 740, 1,834, 752, 960, 1,038
15. 1,188, 1,404, 4,011, 2,740, 5,058, 7,568
16. 8,718, 6,375, 8,288, 23,036, 29,144, 23,435
17. 8,715, 9,308, 10,120, 7,431, 7,172, 10,272
18. 17,880, 21,500, 17,402, 41,274, 19,434, 17,900
19. 819, 76, 7,524, 1,535, 2,415, 1,672
20. 17,332, 4,374, 2,142, 114, 2,064, 21,522

Carrying with Larger Numbers
pages 88–89
1. 5,332, 2,444, 7,790, 28,614, 24,534, 40,291
2. 2,368, 4,482, 1,372, 3,792, 1,537, 4,800
3. 2,072, 1,914, 5,628, 513, 4,032, 1,620
4. 17,250, 20,335, 13,340, 27,676, 12,584, 14,688
5. 31,418, 9,855, 37,855, 11,210, 21,170, 21,875
6. 140,442, 131,008, 118,664, 397,740, 77,418, 344,052
7. 76,284, 34,290, 180,810, 65,274, 66,300, 91,368
8. 115,995, 176,638, 270,115, 275,940, 171,080, 224,110
9. 262,976, 212,725, 788,964, 37,444, 650,650, 155,220
10. 196,174, 119,192, 667,635, 1,150,780, 653,484, 2,820,664

Multiplying Numbers in a Row
page 90
1. 3,616, 43,000, 15,503
2. 23,653, 26,128, 238,502
3. 76,600, 413,040, 202,301
4. 1,540, 13,547, 665,200

Carrying with Dollars and Cents
page 91
1. $46.96, $19.04, $18.13, $20.34, $22.80, $22.80
2. $106.12, $116.62, $84.06, $62.25, $212.80, $148.00
3. $3,187.50, $2,524.50, $12,087.50, $5,962.20, $20,002.00
4. $7.76, $3.36, $4.15
5. $8.76, $16.32, $29.44

Working with Approximations
pages 92–93
1. 130   3. 5,080   5. 22,000   7. 20,000
2. 90   4. 102,000   6. 53,000   8. 63,000

9. 140   12. 5,800   15. 800   18. 7,200
10. 620   13. 50   16. 3,800   19. 60,000
11. 8,800   14. 200   17. 1,200   20. 350,000

Using Approximation in Word Problems
pages 94–95
1. b   3. a   5. c
2. b   4. c   6. a

Solving Word Problems
pages 96–97
1. Approximate: Exact: $1.29
   $1.30   ×   19
   ×   20   11 61
   $26.00   12 9
                            $24.51

2. Approximate: 350
                 + 500
                 850 miles
   Exact: 349
          + 497
          846 miles

3. Approximate: 300
                 − 100
                 200 pounds
   Exact: 298
          − 99
          199 pounds

4. Approximate: 30
                 × 15
                 150
                 30
                 450 miles
   Exact: 29
          × 15
          145
          29
          435 miles

5. Approximate: 400
                   150
                 + 300
                 850 calories
   Exact: 403
           149
         + 296
         848 calories

6. Approximate: $ .50   Exact: $ .49
                 ×   30   ×   31
                 $15.00   .49
                                 14.7
                                 $15.19

7. Approximate: $1.50   Exact: $1.49
                 ×   6   ×   6
                 $9.00 $8.94

176

8. Approximate: $13.00   Exact: $12.89
              + 3.50   + 3.49
              $16.50   **$16.38**

9. Approximate: $7.00   Exact: $6.98
            ×   50   ×   51
          $350.00   6.98
                      349.0
                 $355.98

10. Approximate: 200
            − 100
          100 miles
   Exact: 197
         − 98
      99 miles

11. Approximate: 40
          × 10
       400 miles
   Exact: 39
        × 11
        39
       39
     429 miles

12. Approximate: 20
            10
         + 30
      60 inches
   Exact: 19
         11
       + 29
     59 inches

Multiplication Skills Review
pages 98–99

1. 8, 7, 54, 8, 40, 30
2. 96, 420, 286, 408, 5,313, 88,830
3. 90, 368, 1,692, 1,854, 2,667, 2,688
4. 1,128, 4,042, 2,100, 13,014, 66,270, 106,500
5. $33.60, $40.28, $4,549.00, $41.22, $42.16, $421.60
6. 3,912, 54,000, 7,946, 41,192
7. 30 × 9 = 270
8. 224 × 18 = 4,032
9. 60 × 50 = 3,000
10. $16.24
11. $3.95
12. $15.75
13. $ .25   15. $134.14
    ×   85   ×   48
  $21.25   **$6,438.72**
14. $427
   × 16
  $6,832
16. Approximate: $1.00   Exact: $1.22
           ×   20   ×   19
         $20.00   10.98
                    12.2
                $23.18

17. Approximate: 40   Exact: 41
          × 20   ×19
       800 miles   369
                   41
              779 miles

Chapter 5: Division
Concepts in Division
pages 100–101

1. 9   4. 5   8. $7\overline{)42}$ miles
  72   40
   8   8   9. $10\overline{)30}$ gallons
2. 7
  14   5. $2\overline{)16}$ hours 10. $5\overline{)25}$
   2   6. $6\overline{)\$24}$
3. 4   7. $3\overline{)18}$
  24
   6

Basic Division Facts
pages 102–103

1. 6, 4, 8   8. 3, 4, 8, 6, 3
2. 7, 2, 4   9. 6, 3, 9, 4, 2
3. 2, 8, 9   10. 9, 8, 5, 8
4. 9, 3, 4   11. 5, 2, 7, 9
5. 6, 3, 9   12. $5, 4, 9 coins
6. 5, 7, 7, 3, 9 13. 9, 8, 6 inches
7. 9, 4, 8, 5, 2 14. 4, 7, 3 hours

Dividing by One Digit
page 104

1. 21, 13, 24, 12, 11, 21, 14
2. 11, 11, 11, 23, 31, 12, 11
3. 211, 214, 231, 111, 141, 121
4. 112, 212, 221, 323, 421, 111
5. 2,112, 2,412, 3,213, 31,242, 32,231, 21,122

Dividing into Zero
page 105

1. 20, 20, 20, 320, 320, 410
2. 200, 100, 200, 200, 300, 400
3. 201, 304, 302, 202, 101, 203
4. 3,001, 2,002, 4,001, 1,001, 3,002, 1,002

Dividing into a Smaller Digit
page 106

1. 51, 53, 31, 42, 931, 712
2. 81, 51, 54, 41, 512, 211
3. 52, 81, 50, 101, 52, 20
4. 20, 100, 201, 902, 530, 502

Using Zero as a Place Holder
page 107

1. 204, 103, 309, 309, 102, 106
2. 2,304, 2,104, 1,208, 1,105, 1,104, 1,102
3. 121, 203, 60, 50, 102, 300
4. 40, 102, 51, 80, 302, 205
5. 2,013, 2,010, 201, 600, 930, 501

Remainders in Division
pages 108–109

1. 1r3, 1r1, 2r4, 2r5, 7r3, 7r2
2. 1r3, 1r4, 1r1, 2r1, 2r1, 1r2
3. 3r2, 7r2, 6r8, 6r3, 7r3, 7r2

177

4.

$$\begin{array}{r} 4\ r\ 3 \\ 5\overline{)23} \\ \underline{20} \\ 3 \end{array}\qquad \begin{array}{r} 5\ r\ 6 \\ 7\overline{)41} \\ \underline{35} \\ 6 \end{array}\qquad \begin{array}{r} 6\ r\ 7 \\ 9\overline{)61} \\ \underline{54} \\ 7 \end{array}$$

5.

$$\begin{array}{r} 7\ r\ 8 \\ 10\overline{)78} \\ \underline{70} \\ 8 \end{array}\qquad \begin{array}{r} 8\ r\ 3 \\ 6\overline{)51} \\ \underline{48} \\ 3 \end{array}\qquad \begin{array}{r} 9 \\ 8\overline{)72} \\ \underline{72} \\ 0 \end{array}$$

Introducing Long Division
pages 110–111

1. 14, 17, 24, 23, 17, 87
2. 18, 15, 13, 17, 12, 15
3. 34, 93, 79, 59, 62, 88
4. 85, 15, 22, 69, 75, 45
5. 18r3, 13r2, 11r5, 26r1, 52r1, 89r1
6. 12r2, 12r5, 18r3, 11r7, 28r1, 27r2
7. 94r4, 87r1, 45r2, 98r6, 84r3, 87r1

Dividing Dollars and Cents
page 112

1. $0.38, $0.72, $0.34, $0.81, $0.55, $0.42
2. $0.08, $0.07, $0.10, $0.08, $0.09, $0.09
3. $1.08, $0.47, $0.04

Division Word Problems
pages 113–115

1. **Key word:** each

$$\begin{array}{r} 13\ \text{cards} \\ 4\overline{)52} \\ \underline{4} \\ 12 \\ \underline{12} \\ 0 \end{array}$$

2. **Key word:** single

$$\begin{array}{r} \$0.39 \\ 3\overline{)\$1.17} \\ \underline{9} \\ 27 \\ \underline{27} \\ 0 \end{array}$$

3. **Key words:** share, each, equally

$$\begin{array}{r} \$\ 800,100 \\ 6\overline{)\$4,800,600} \\ \underline{4,800,600} \\ 0 \end{array}$$

4. **Key words:** average, each

$$\begin{array}{r} 22\ \text{hamburgers} \\ 8\overline{)176} \\ \underline{16} \\ 16 \\ \underline{16} \\ 0 \end{array}$$

5. **Key words:** divided, each

$$\begin{array}{r} 4\ \text{pieces} \\ 4\overline{)16} \\ \underline{16} \\ 0 \end{array}$$

6. **Key word:** every

$$\begin{array}{r} \$\ 7.10 \\ 9\overline{)\$63.90} \\ \underline{63} \\ 0\ 90 \\ \underline{90} \\ 0 \end{array}$$

7. **Key words:** cut, each, equal pieces

$$\begin{array}{r} 49\ \text{inches} \\ 3\overline{)147} \\ \underline{12} \\ 27 \\ \underline{27} \\ 0 \end{array}$$

8. **Key word:** split

$$\begin{array}{r} \$\ 4.10 \\ 5\overline{)\$20.50} \\ \underline{20} \\ 0\ 50 \\ \underline{50} \\ 0 \end{array}$$

9.

$$\begin{array}{r} 4\ \text{dresses} \\ 2\overline{)9} \\ \underline{8} \\ 1 \end{array}$$

10.

$$\begin{array}{r} 14\ \text{weeks} \\ \$9\overline{)\$126} \\ \underline{9} \\ 36 \\ \underline{36} \\ 0 \end{array}$$

11.

$$\begin{array}{r} 6\ r\ 3 \\ 6\overline{)39} \\ \underline{36} \\ 3 \end{array}$$

(Amy will need 7 boxes to pack all the records.)

12.

$$\begin{array}{r} 17\ \text{months} \\ 3\overline{)51} \\ \underline{3} \\ 21 \\ \underline{21} \\ 0 \end{array}$$

More About Long Division
pages 116–117

1. 122, 139, 248, 265, 926
2. 143, 274, 268, 117, 228
3. 738, 324, 869, 196, 634
4. 163r1, 146r2, 132r2, 143r2, 418r3
5. 119r1, 275r2, 185r3 462r4, 532r2

Dividing by Two Digits
pages 118–119

1. 2, 2, 2, 4, 2
2. 4, 3, 3, 5, 3
3. 5, 5, 4, 3, 4
4. 4, 6r6, 5r1, 3, 3
5. 8, 5, 2r5, 6, 3r4
6. 5, 8, 5, 7, 4
7. 8r2, 5, 3r9, 6r6, 7
8. 7, 7r4, 5, 8r6, 5

Dividing Larger Numbers
pages 120–122

1. 17, 38, 35, 34, 42   2. 27, 18, 57, 19, 21

3. 41, 31, 28, 22, 50   8. 6, 3, 6, 4
4. 125, 126, 234, 241   9. 5, 4, 8, 4
5. 127, 213, 125, 234   10. 17, 21, 24, 13
6. 215, 246, 321, 214   11. 78, 54, 61, 58
7. 4, 5, 18, 93

Dividing Numbers in a Row
page 123
1. 22r12, 6, 14, 22r4
2. 23, 22, 38, 36r11
3. 135, 128r26, 4, 4r27
4. 32, 51r128, 71

Using Division to Check Multiplication
page 124
2. 520
4. $19.08
6. 9,579
8. 87,584

Deciding When to Multiply or Divide
pages 125–127
1. **Multiplied:** $ 0.79
           ×   150
           39.50
           79.
           $118.50

2. **Divided:**   **165 days**
        $2)$330
              2
              13
              12
              10
              10
              0

3. **Multiplied:** $ 1.65
           ×   9
           $14.85

4. **Divided:**   **23 miles**
        7)161
          14
          21
          21
          0

5. **Multiplied:**   168
          ×   4
          672 quarts

6. b
7. a
8. b
9. b
10. a
11. a
12. b
13. a

Solving Word Problems
pages 128–131

1.   **78 words**   2.   **$0.79**
    4)312   24)$18.96
      28   16 8
      32   2 16
      32   2 16
      0   0

3. $285   4.   **4 pounds**
    ×   12   36)144
    570   144
    2 85   0
    $3,420

5.   **36 minutes**   6.   **$ 89**
    240)8640   4)$356
       720   32
       1440   36
       1440   36
       0   0

7.   37
    × 16
    222
    37
    592 miles

8.   **33 r 25**
    75)2500
       225
       250
       225
       25

(Jenny must buy **34** cartons in order to have enough cups.)

9.   1,238   10. $13.49
    ×   13   ×   47
    3 714   94.43
    12 38   539.6
    16,094 records   **$634.03**

11.   **9.91**
    $2)$19.82
       18
       1 82
       1 8
       2
       2
       0

(Grace could buy **9** bottles of shampoo.)

12.   **308 words**
    234)72072
       702
       1872
       1872
       0

13.   **73 words per minute**
    6)438
      42
      18
      18
      0

14. $935
× 6
$5,610

15. $1,156
× 12
2 312
11 56
$13,872

16. $237.71
× 24
950.84
4 754.2
$5,705.04

17. **$ 240.20**
24) $5764.80
48
96
96
04 8
4 8
00

18. $540
× 52
1 080
27 00
$28,080

19. **$ 284**
52) $14768
104
436
416
208
208
0

20. $4.75
× 18
38 00
47 5
$85.50

Division Skills Review
pages 132–133

1. 5, 9, 8, 9, 3, 7
2. 21, 312, 340, 61, 1,020, 200
3. 6r1, 6r4, 15, 18, 53r6, 75
4. 45r5, 229, 21, 24, 25, 30r5
5. $.30, $.33, $.32, $2.07, $1.50
6. 68, 16r1, 223, 1,045

7. **12**
5) 60
5
10
10
0

8. **6**
15) 90
90
0

9. **$ 3**
12) $36
36
0

10. **.19**
4) .76
4
36
36
0

21. **147 boxes**
128) 18816
128
601
512
896
896
0

22. **91 boxes**
48) 4368
432
48
48
0

23. 144
× 98
1 152
12 96
14,112 cans

24. 100
× 48
800
4 00
4,800 cans

11. **$0.60**
8) $4.80
4 8
00

12. **$ 2.10**
14) $29.40
28
1 4
1 4
00

15. **7 r 2**
6) 44
42
2
(She can make **7** complete statues.)

16. **22 pieces**
$24) $528
48
48
48
0

17. **$ 39.93**
6) $239.58
18
59
54
5 5
5 4
18
18
0

13. **$0.46**
3) $1.38
1 2
18
18
0

14. **46 minutes**
65) 2990
260
390
390
0

Chapter 6: Special Topics
Introduction to Multi-Step Word Problems
pages 134–139

1. STEP 1: $3.58
1.89
+ 2.79
$8.26
   STEP 2: $20.00
− 8.26
$11.74

2. STEP 1:   121
× 5
605 Monday – Friday papers
   STEP 2:   605
+257
862 total papers

3. STEP 1:   84
−12
72 cookies
   STEP 2:   **3 cookies each**
24) 72
72
0

4. STEP 1: **Add.** $12.75
 4.30
 9.45
 $26.50 total cost of dinner
 STEP 2: **Divide.** **$ 5.30 each person's**
 5)$26.50 **share**
 25
 1 5
 1 5
 00

5. STEP 1: **Add.** $0.79
 1.98
 + 0.35
 $3.12 cost of 1 package
 STEP 2: **Multiply.** $3.12
 × 23
 9.36
 62.4
 $71.76

6. STEP 1: **Multiply.** 15
 ×24
 60
 30
 360 quarts
 STEP 2: **Divide.** **72 tune-ups**
 5)360
 35
 10
 10
 0

7. STEP 1: $28.50
 13.65
 + 0.83
 $42.98 total cost of parts
 STEP 2: $75.40
 − 42.98
 $32.42 labor cost

8. STEP 1: 8
 +6
 14 boxes per trip
 STEP 2: 14
 × 8
 112 total boxes moved

9. STEP 1: $ 85
 285
 + 98
 $468 total expenses
 STEP 2: **$156 each club's share**
 3)$468
 3
 16
 15
 18
 18
 0

10. STEP 1: 55
 ×30
 1,650 balloons given away

11. STEP 2: 1,650
 + 245
 1,895 total balloons

11. **Solution sentence:** change = $100
 minus total cost of dresses
 STEP 1: $34.95
 27.50
 + 19.95
 $82.40
 STEP 2: $100.00
 − 82.40
 $ 17.60 change

12. **Solution sentence:** total cost of
 supplies = cost of manuals *plus* cost of
 first aid kit
 STEP 1: $6.49
 × 17
 45.43
 64.9
 $110.33 cost of manuals
 STEP 2: $110.33
 + 17.50
 $127.83 total cost of supplies

13. **Solution sentence:** number of people
 served = 95 times 4 (days) *plus* 165
 STEP 1: 95
 × 4
 380 served in 4 days
 STEP 2: 380
 +165
 545 served in 5 days

14. **Solution sentence:** each brother's share
 of rent = total rental cost *divided by* 3
 STEP 1: $27.84
 × 5
 $139.20 total rental cost
 STEP 2: **$ 46.40 each brother's share**
 3)$139.20
 12
 19
 18
 12
 12
 00

15. **Solution sentence:** miles on Toyota's
 indicator = miles per month times 12
 plus 32,696
 STEP 1: 1,350
 × 12
 2 700
 13 50
 16,200 miles
 STEP 2: 16,200
 +32,696
 48,896 miles on indicator

16. **Solution sentence:** amount of check =
 cost of items *plus* $5

STEP 1: $2.89
           3.49
        + 0.79
          $7.17
STEP 2: $ 7.17
        + 5.00
       $12.17 amount of check

17. **Solution sentence:** number of voters
were neither Republicans nor Democrats
= 14,725 *minus* number of Republicans
and number of Democrats combined
STEP 1:   6,539
          + 4,891
            11,430 Republicans and
                   Democrats combined
STEP 2:   14,725
          − 11,430
            **3,295 number of voters
                    neither Republican
                    nor Democrat**

English and Metric Measurement
page 141
1. longer
2. heavier
3. larger
4. less than
5. millimeter
6. milligram
7. milliliter
8. 10 meters
9. 5 kilograms
10. 3 liters

Smaller Units in a Larger Unit
page 142
1. 1 minute = 60 seconds
   1 hour = 60 minutes
        60
      × 60
   3,600 seconds in an hour

2. 1 foot = 12 inches
   1 mile = 5,280 feet
     5,280
   ×   12
    10 560
    52 80
   63,360 inches in a mile

3. 1 meter = 100 centimeters
   1 kilometer = 1,000 meters
        1,000
   ×   100
   100,000 centimeters in a kilometer

4. 1 kilogram = 1,000 grams
   1 metric ton = 1,000 kilograms
         1,000
   ×   1,000
   1,000,000 grams in a metric ton

5. 1 pint = 16 ounces
   1 quart = 2 pints
       16
   ×   2
   32 ounces in a liquid quart

6. 1 meter = 1,000 millimeters
   1 kilometer = 1,000 meters
         1,000
   ×   1,000
   1,000,000 millimeters in a kilometer

Changing from One Unit to Another
pages 143–144
1. 84, 24, 120
2. 48, 5,000, 2,000
3. 6, 2,000, 240
4. 3, 3, 6
5. 3, 3, 2
6. 8, 2, 5
7. 2 ft. 5 in., 4 yds. 2 ft.
8. 1 hr. 15 min., 2 lbs. 15 oz.
9. 2 m 40 cm, 2 cm 6 mm
10. 2 kg 500 g, 3 L 400 ml

Simplifying Mixed Units
page 145
1. 7 ft. 7 in., 24 cm 4 mm
2. 13 yd. 1 ft., 4 lb. 1 oz.
3. 16 m 50 cm, 6 qt. 1 pt.
4. 7 L 750 ml, 7 min. 7 sec.
5. 10 ft. 8 in., 5 kg 300 g

Adding Measurement Units
pages 146–147
1. 7 yd. 1 ft., 9 yd., 17 ft. 3 in.,
   6 mi. 190 yd.
2. 13 cm 2 mm, 9 cm 3 mm, 38 m 21 cm,
   78 km 250 m
3. 8 lb. 5 oz., 10 lb. 7 oz., 2 lb. 13 oz.,
   4 T. 1975 lb.
4. 14 kg 650 g, 5 kg 235 g, 22 g 25 mg,
   4 t 449 kg
5. 8 qt., 8 qt., 5 pt. 1 oz., 3 gal. 1 qt.
6. 4 L 550 ml, 6 L 250 ml, 26 L 125 ml,
   4 kl 250 L
7. 6 hr. 20 min., 4 hr. 8 min.,
   13 min. 4 sec., 4 da. 1 hr.
8.   5 yd. 2 ft.
      4 yd. 1 ft.
   + 6 yd. 2 ft.
     15 yd. 5 ft. = **16 yd. 2 ft.**
9.   4 km   800 m
      5 km   400 m
   + 6 km   100 m
     15 km 1300 m = **16 km   300 m**
10.   2:25
    + 3:45
      5:70 = **6:10**

11. 1 L   500 ml
         350 ml
     +   500 ml
     1 L 1350 ml = **2 L 350 ml**

12. 2 lb. 7 oz.
     1 lb. 3 oz.
     +   13 oz.
     3 lb. 23 oz. = **4 lb. 7 oz.**

13. 10 kg. 750 g.
     ×   3
     30 kg. 2250 g. = **32 kg. 250 g.**

Subtracting Units of Measurement
pages 148–149

1. 2 ft. 6 in., 3 ft. 11 in., 1 yd. 2 ft.,
   1,110 yd.
2. 2 cm 6 mm, 3 cm 4 mm, 3 m 67 cm,
   2 km 825 m
3. 3 lb. 15 oz., 1 lb. 13 oz., 14 oz., 1,205 lb.
4. 3 kg 275 g, 1 kg 775 g, 2 g 400 mg,
   1 t 950 kg
5. 1 qt. 1 pt., 2 qt. 1 pt., 1 pt. 13 oz., 3 qt.
6. 625 ml, 1 L 815 ml, 505 ml, 1 kl 575 L
7. 55 min., 1 hr. 40 min., 2 min. 55 sec.,
   1 da. 18 hr.
8.   1 m 23 cm   10.   57 yd. 1 ft.
     −   98 cm   − 42 yd. 2 ft.
         25 cm.   **14 yd. 2 ft.**
9. 3 hr. 15 min.   11.   1 L   200 ml
    − 1 hr. 45 min.   −   350 ml
    1 hr. 30 min.   **850 ml**

Multiplying Measurement Units
pages 150–151

1. 22 ft. 8 in., 11 ft. 3 in., 26 yd. 2 ft., 5 mi.
2. 11 m 55 cm, 11 m 4 cm, 11 cm 2 mm,
   38 km 200 m
3. 7 lb. 8 oz., 7 lb. 6 oz., 15 lb., 4 T. 475 lb.
4. 17 kg 200 g, 20 kg 400 g, 26 g 125 mg,
   5 t 500 kg
5. 16 qt. 1 pt., 17 qt. 1 pt., 4 pt. 5 oz.,
   5 gal. 2 qt.
6. 7 L 50 ml, 3 L 150 ml, 9 L 200 ml,
   10 kl 950 L
7. 5 hr. 15 min., 13 hr. 40 min., 34 min.,
   3 yr. 35 da.
8.   6 ft. 8 in.
     ×   4
     24 ft. 32 in. = **26 ft. 8 in.**
9.   1 lb. 3 oz.
     ×   8
     8 lb. 24 oz. = **9 lb. 8 oz.**
10.   1 kg 250 g
      ×   12
      12 kg 3000 g = **15 kg**
11. 1 pt. 6 oz.
     ×   5
     5 pt. 30 oz. = **6 pt. 14 oz.**
12. 375 L × 12 = 4,500 L = **4 kl 500 L**

13.   4 cm 3 mm
    ×   6
    24 cm 18 mm = **25 cm 8 mm**

Dividing Measurement Units
pages 152–153

1. 1 ft. 9 in., 1 ft. 6 in., 3 yd. 2 ft.,
   1 mi. 598 yd.
2. 2 cm 4 mm, 2 cm 5 mm, 2 m 71 cm,
   3 km 744 m
3. 2 lb. 10 oz., 2 lb. 7 oz., 1 T. 1,300 lb.,
   2 kg. 210 g.
4. 1 gal. 3 qt., 2 gal. 3 qt., 1 pt. 8 oz., 6 oz.
5. 1 L 320 ml, 1 L 825 ml, 625 ml, 1 kl 510 L
6. 1 hr. 40 min., 1 hr. 30 min.,
   2 min. 30 sec., 1 da. 16 hr.

7.   **1 ft.   4 in.**   9.   **27 kg   500 g**
    5) 6 ft.   8 in.   3) 82 kg   500 g
       5   81
    1 ft. = 12 in.   1 kg = 1000 g
          20   1500
          20   1500
           0   00

8.   **4 lb.   3 oz.**   10.   **1 hr. 30 min.**
    7) 29 lb.   5 oz.   6) 9 hrs.
       28   6
    1 lb. = 16 oz.   3 hrs. = 180 min.
          21   180
          21   00
           0

Learning About Perimeter
pages 154–155

1. 20 yd. 2 ft.   3. 12 cm 2 mm
2. 340 m 50 cm   4. 11 ft. 6 in.
5.   21
     × 4
     84 yards
6. STEP 1:   32   20
            × 2   × 2
              64   40
   STEP 2:   64
            + 40
            104 inches
7.   9 cm 4 mm
     12 cm 5 mm
   + 14 cm
   35 cm 9 mm
8. STEP 1:   450   285
           ×   2   ×   2
           900 yards   570 yards
   STEP 2:   900
           + 570
           1,470 yards
   STEP 3:   1,470 yards
           ×   5
           7,350 total yards

Becoming Familiar with Area
pages 156–157

1. 8 sq. feet
2. 15 sq. meters

3. 5 sq. yards
4. 6 sq. centimeters
5. 6 sq. feet
6. 11 sq. miles
7.   4 yards
    × 3 yards
    12 sq. yards

8. $12.50
  ×   12
  25.00
 125.0
 $150.00

9. STEP 1: A =   10 feet
            × 7 feet
            70 sq. feet
   STEP 2:   **70 tiles**
        1 sq. foot$\overline{)}$70 sq. feet

10.   120 yards
   × 60 yards
   7,200 sq. yards

11. STEP 1:   50 meters
         × 35 meters
         250
       1 50
     1,750 sq. meters
   STEP 2:   **17 r 50**
    100 sq. meters$\overline{)}$1750 sq. meters
             100
             750
             700
              50

(Guy will need **18** bags to do the job.)

Becoming Familiar with Volume
pages 158–159
1. 16 cubic yards
2. 18 cubic meters
3. 15 cubic inches
4. STEP 1:   4 yards
         × 3 yards
         12 sq. yards
   STEP 2:   12 sq. yards
          × 3 yards
          36 cubic yards
5. STEP 1:   3 feet
         × 2 feet
         6 sq. feet
   STEP 2:   6 sq. feet
         × 2 feet
         12 cubic feet
6. STEP 1:   6 feet
         × 5 feet
         30 sq. feet
   STEP 2:   30 sq. feet
         × 1 foot
         30 cubic feet
   STEP 3:   30 cubic feet
         × 62 pounds
         60
      1 80
     1,860 pounds

7. STEP 1: Volume of block
        = 3 × 3 × 3
        = 27 cubic inches
  STEP 2: Volume of shipping box =
    a.   24 inches
       × 18 inches
       192
       24
      432 sq. inches
    b.   432 sq. inches
      × 12 inches
      864
     4 32
    5,184 cubic inches
  STEP 3.   **192 blocks will fit**
    27$\overline{)}$5184
      27
     248
     243
      54
      54
       0

Finding an Average
pages 160–161
1. STEP 1:   124
       120
     + 125
     369 pounds
   STEP 2:   **123 pounds**
    3$\overline{)}$369
     3
    06
     6
    09
     9
     0
2. STEP 1:   480
       362
       412
     + 290
     1,544 miles
   STEP 2:   **386 miles**
    4$\overline{)}$1544
    12
    34
    32
    24
    24
     0
3. STEP 1: 82   STEP 2:   **82**
       73   4$\overline{)}$328
       87   32
     + 86   08
     328   8
                0

4. STEP 1: $318   STEP 2: **$312**

```
    $318              STEP 2:  $312
     289                     3)$936
   +  329                       9
    $936                        03
                                 3
                                 06
                                  6
                                  0
```

5. STEP 1:

```
    $3.60             STEP 2:   $ 2.34
     2.85                     5)$11.70
     2.25                       10
     1.75                       1 7
   +  1.25                      1 5
    $11.70                        20
                                  20
                                   0
```

6. STEP 1:

```
    2,112             STEP 2:   2042 fans
    2,080                     4)8168
    1,984                       8
  + 1,992                       016
    8,168 fans                   16
                                 08
                                  8
                                  0
```

7. STEP 1:

```
    1 hr.  45 min.
    2 hr.  10 min.
  + 1 hr.  53 min.
    4 hr. 108 min.
```

STEP 2: **1 hr.   56 min.**

```
    3) 4 hr.  108 min.
       3
       1 hr. = 60 min.
              168 min.
               15
               18
               18
                0
```

8. STEP 1:

```
    $0.84             STEP 2:   $0.81
     0.79                     5)$4.05
     0.81                       4 0
     0.79                       05
   +  0.82                        5
    $4.05                         0
```

9. STEP 1:

```
    6 lb.  9 oz.
    7 lb. 11 oz.
  + 6 lb.  3 oz.
   19 lb. 23 oz.
```

STEP 2: **6 lb.   13 oz.**

```
    3) 19 lb.   23 oz.
       18
       1 lb. = 16 oz.
               39 oz.
                3
                09
                 9
                 0
```

10. STEP 1:

	1st week	2nd week
	1,800	1,925
	1,750	2,100
	1,770	1,970
	1,810	1,975
	1,800	1,900
	1,780	2,080
	+1,820	+1,980
	12,530	13,930

STEP 2: **1790**   **1990**

```
    7) 12530      7) 13930
       7             7
       55            69
       49            63
       63            63
       63            63
       00            00
```

Squares, Cubes, and Square Roots
pages 162–163

1. 4^2 four squared
2. 5^2 five squared
3. 10^2 ten squared
4. 25^2 twenty-five squared
5. 36^2 thirty-six squared
6. 3^3 three cubed
7. 8^3 eight cubed
8. 10^3 ten cubed
9. 25^3 twenty-five cubed
10. 32^3 thirty-two cubed
11. 9

12. 49
13. 81
14. 144
15. 225
16. 400
17. 27
18. 125
19. 343
20. 729
21. 1,728
22. 8,000

What Is a Square Root?
page 164

1. $4 = \sqrt{16}$
2. $10 = \sqrt{100}$
3. $9 = \sqrt{81}$
4. $12 = \sqrt{144}$
5. 8
6. 2
7. 11
8. 5
9. 7
10. 15
11. 1
12. 14
13. 6

Post-Test
pages 165–167

1. 8
2. 7
3. 9
4. 8
5. 39
6. 66
7. 87
8. 124
9. $9.11
10. 1,097
11. 6,775
12. 7,412
13. 3
14. 8
15. 33
16. 17
17. 786
18. $13.39
19. 612
20. 152

21. 351
22. 422
23. $46.37
24. 40
25. 106
26. 276
27. 39,483
28. 1,560
29. 106,530
30. 376
31. 7,154
32. 411,546
33. 4,484
34. 4,508,000
35. 5
36. 32
37. 2,123
38. 30
39. 209
40. 903

41. 6 r 3
42. 13 r 3
43. $.33
44. 30
45. 23
46. 213
47. 8
48. 15
49. 664 r 5
50. $.50
    .25
    .10
   +.01
   $.86

answer is (c)

51.　936
　　 −795
　　 141 miles farther
　　 answer is (a)

52.　$23.49
　　 19.95
　　 + 19.95
　　 $63.39
　　 answer is (b)

53.　$1.49
　　 × 21
　　 1 49
　　 29 8
　　 $31.29
　　 answer is (d)

54.　145
　　 −131
　　 14 pounds
　　 answer is (c)

55.　 58 words per minute
　　 6)348
　　 30
　　 48
　　 48
　　 0
　　 answer is (a)

56. STEP 1:　$0.99
　　　　　 × 6 screwdrivers
　　　　　 $5.94 screwdrivers cost
　　 STEP 2:　$ 5.94
　　　　　 + 39.95
　　　　　 $45.89 cost of tools
　　 STEP 3:　$50.00
　　　　　 − 45.89
　　　　　 $ 4.11 change
　　　　　 answer is (b)

57.　 2 qt. 1 pt.
　　 × 6
　　 12 qt. 6 pt. = 3 gal. 3 qt.
　　 answer is (b)

58. STEP 1:　40 feet
　　　　　 × 32 feet
　　　　　 80
　　　　　 1 20
　　　　　 1,280 square feet
　　 STEP 2:　 20 rolls
　　　　　 64)1,280 square feet
　　　　　 1 28
　　　　　 00
　　 STEP 3:　$25.50
　　　　　 × 20 rolls
　　　　　 $510.00
　　　　　 answer is (e)

59. STEP 1:　144
　　　　　 151
　　　　　 167
　　　　　 +146
　　　　　 608
　　 STEP 2:　 152 average score
　　　　　 4)608
　　　　　 4
　　　　　 20
　　　　　 20
　　　　　 08
　　　　　 8
　　　　　 0
　　　　　 answer is (e)

60. STEP 1:　4 feet
　　　　　 ×2 feet
　　　　　 8 square feet
　　 STEP 2:　8 square feet
　　　　　 ×3 feet
　　　　　 24 cubic feet
　　　　　 answer is (b)